CHEMISTRY RESEARCH AND APPLICATIONS SERIES

CYCLIC β-KETO ESTERS: SYNTHESIS AND REACTIONS

CHEMISTRY RESEARCH AND APPLICATIONS SERIES

Applied Electrochemistry
Vijay G. Singh (Editor)
2009. ISBN: 978-1-60876-208-8

Handbook on Mass Spectrometry: Instrumentation, Data and Analysis, and Applications
J. K. Lang (Editor)
2009. ISBN: 978-1-60741-580-0

Handbook of Inorganic Chemistry Research
Desiree A. Morrison (Editor)
2010. ISBN: 978-1-61668-010-7

Handbook of Inorganic Chemistry Research
Desiree A. Morrison (Editor)
2010. ISBN: 978-1-61668-712-0 (Online Book)

Solid State Electrochemistry
Thomas G. Willard (Editor)
2010. ISBN: 978-1-60876-429-7

Mathematical Chemistry
W. I. Hong (Editor)
2010. ISBN: 978-1-60876-894-3

Mathematical Chemistry
W. I. Hong (Editor)
2010. ISBN: 978-1-61668-440-2 (Online Book)

Physical Organic Chemistry: New Developments
Karl T. Burley (Editor)
2010. ISBN: 978-1-61668-435-8

Physical Organic Chemistry: New Developments
Karl T. Burley (Editor)
2010. ISBN: 978-1-61668-469-3 (Online Book)

Chemical Sensors: Properties, Performance and Applications
Ronald V. Harrison (Editor)
2010. ISBN: 978-1-60741-897-9

Macrocyclic Chemistry: New Research Developments
Dániel W. Fitzpatrick and Henry J. Ulrich (Editors)
2010. ISBN: 978-1-60876-896-7

Electrolysis: Theory, Types and Applications
Shing Kuai and Ji Meng (Editors)
2010. ISBN: 978-1-60876-619-2

Energetic Materials:
Chemistry, Hazards and Environmental Aspects
Jake R. Howell and Timothy E. Fletcher (Editors)
2010. ISBN: 978-1-60876-267-5

Chemical Crystallography
Bryan L. Connelly (Editor)
2010. ISBN: 978-1-60876-281-1

Chemical Crystallography
Bryan L. Connelly (Editor)
2010. ISBN: 978-1-61668-513-3 (Online Book)

Heterocyclic Compounds: Synthesis, Properties and Applications
Kristian Nylund and Peder Johansson (Editors)
2010. ISBN: 978-1-60876-368-9

Influence of the Solvents on Some Radical Reactions
Gennady E. Zaikov, Roman G. Makitra, Galina G. Midyana and Lyubov.I Bazylyak
2010. ISBN: 978-1-60876-635-2

Dzhemilev Reaction in Organic and Organometallic Synthesis
Vladimir A.D'yakonov
2010. ISBN: 978-1-60876-683-3

Analytical Chemistry of Cadmium: Sample Pre-Treatment and Determination Methods
Antonio Moreda-Piñeiro and Jorge Moreda-Piñeiro
2010. ISBN: 978-1-60876-808-0

Binary Aqueous and CO2 Containing Mixtures and the Krichevskii Parameter
Aziz I. Abdulagatov, Ilmutdin M. Abdulagatov and Gennadii V. Stepanov
2010. ISBN: 978-1-60876-990-2

Advances in Adsorption Technology
Bidyut Baran Saha and Kim Choon Ng (Editors)
2010. ISBN: 978-1-60876-833-2

Electrochemical Oxidation and Corrosion of Metals
Elena P. Grishina and Andrew V. Noskov
2010. ISBN: 978-1-61668-329-0

Electrochemical Oxidation and Corrosion of Metals
Elena P. Grishina and Andrew V. Noskov
2010. ISBN: 978-1-61668-329-0 (Online Book)

Modification and Preparation of Membrane in Supercritical Carbon Dioxide
Guang-Ming Qiu, Rui Tian, Yang Qiu and You-Yi Xu
2010. ISBN: 978-1-60876-905-6

Thermostable Polycyanurates: Synthesis, Modification, Structure and Properties
Alexander Fainleib (Editor)
2010. ISBN: 978-1-60876-907-0

Combustion Synthesis of Advanced Materials
B. B. Khina
2010. ISBN: 978-1-60876-977-3

Structure and Properties of Particulate-Filled Polymer Composites: The Fractal Analysis
G. V. Kozlov, Y. G. Yanovskii and G. E. Zaikov
2010. ISBN: 978-1-60876-999-5

Information Origins of the Chemical Bond
Roman F. Nalewajski
2010. ISBN: 978-1-61668-305-4

Cyclic β-Keto Esters: Synthesis and Reactions
M.A. Metwally and E. G. Sadek
2010. ISBN: 978-1-61668-282-8

Wet Electrochemical Detection of Organic Impurities
F. Manea, C. Radovan, S. Picken and J. Schoonman
2010. ISBN: 978-1-61668-661-1

Wet Electrochemical Detection of Organic Impurities
F. Manea, C. Radovan, S. Picken and J. Schoonman
2010. ISBN: 978-1-61668-491-4 (Online Book)

Quantum Frontiers of Atoms and Molecules
Mihai V. Putz (Editor)
2010.ISBN: 978-1-61668-158-6

Molecular Symmetry and Fuzzy Symmetry
Xuezhuang Zhao
2010. ISBN: 978-1-61668-528-7

Molecular Symmetry and Fuzzy Symmetry
Xuezhuang Zhao
2010. ISBN: 978-1-61668-375-7 (Online Book)

Molybdenum and Tungsten Cofactor Model Chemistry
Carola Schulzke, Prinson P. Samuel
2010. ISBN: 978-1-61668-750-2

Molybdenum and Tungsten Cofactor Model Chemistry
Carola Schulzke, Prinson P. Samuel
2010. ISBN: 978-1-61668-828-8 (Online Book)

Rock Chemistry
Basilio Macías and Fidel Guajardo (Editors)
2010. ISBN: 978-1-60876-563-8

Boron Hydrides, High Potential Hydrogen Storage Materials
Umit B. Demirci and Philippe Miele (Editors)
2010. ISBN: 978-1-61668--362-7

Tetraazacyclotetradecane Species as Models of the Polyazacrown Macrocycles
Ryszard B. Nazarski
2010. ISBN: 978-1-61668-487-7

Chemical Reactions in Gas, Liquid and Solid Phases: Synthesis, Properties and Application
G.E. Zaikov and R.M. Kozlowski
2010. ISBN: 978-1-61668-671-0

CHEMISTRY RESEARCH AND APPLICATIONS SERIES

CYCLIC β-KETO ESTERS: SYNTHESIS AND REACTIONS

M. A. METWALLY
AND
E. G. SADEK

Nova Science Publishers, Inc.
New York

For permission to use material from this book please contact us:
Telephone 631-231-7269; Fax 631-231-8175
Web Site: http://www.novapublishers.com

LIBRARY OF CONGRESS CATALOGING-IN-PUBLICATION DATA

Metwally, M.A.
Cyclic beta-keto esters : synthesis and reactions / M.A. Metwally and E. G. Sadek.
p. cm.
ISBN 978-1-61668-282-8 (softcover)
1. Ketones. 2. Esters. 3. Organic cyclic compounds. I. Sadek, E. G. II. Title.
QD305.K2.M58 2009
547'.638--dc22

2009050559

Published by Nova Science Publishers, Inc. ✦ New York

CONTENTS

PREFACE

The chemistry of cyclic β-keto esters has received considerable attention because they serve as highly important intermediates in many organic synthesis.The wide application of these compounds in organic synthesis, depends chiefly on the reactivity of the methylene group, and the ready hydrolysis of the ester group, and its substituted derivatives to carboxlic acids and ketones. The syntheses and reactions of cyclic β-keto esters are reviewed in a formal way.

Chapter 1

SYNTHESIS

1. CARBOXYLATION OF KETONES

a) With solid carbon dioxide, followed by esterification of the product, e.g. in the cyclehexanone series [1]:

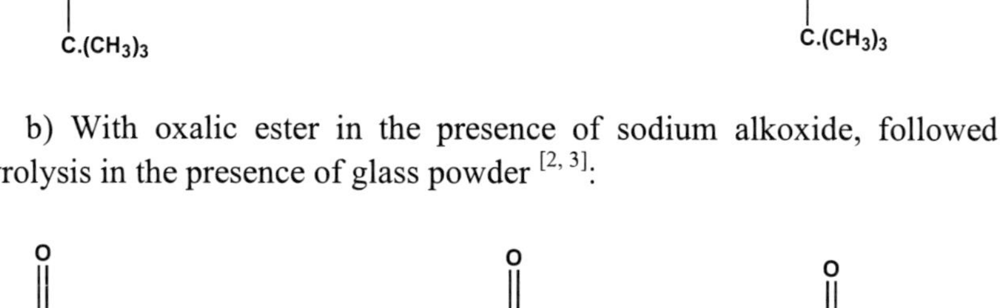

b) With oxalic ester in the presence of sodium alkoxide, followed by pyrolysis in the presence of glass powder [2, 3]:

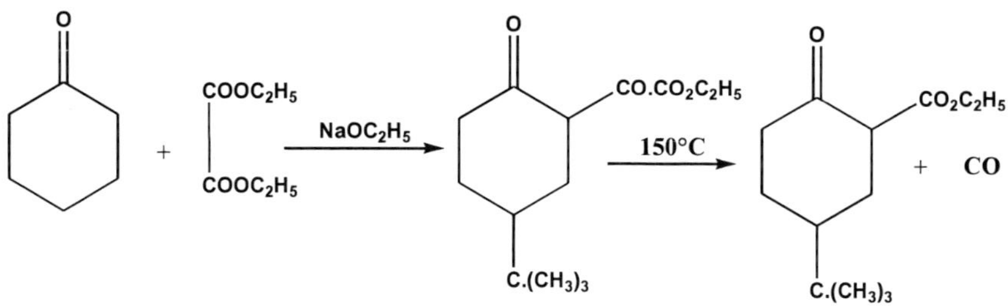

2. Intramolecular Condensation and Cyclization of Six-Membered Rings

Six-membered rings may also be formed by intramolecular condensation and cyclization, example of which may be represented as follows:

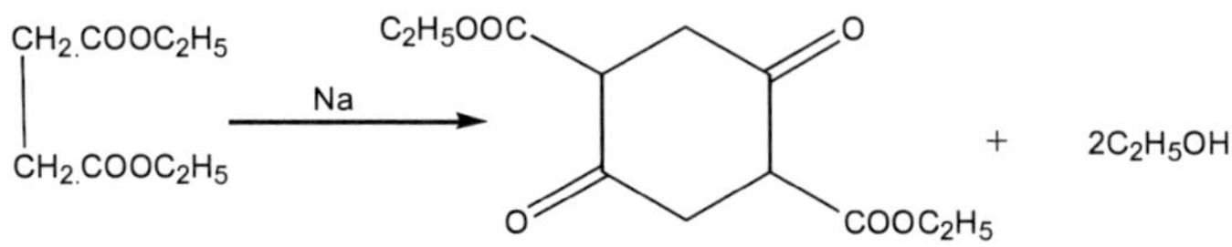

3. Isomerization

Ethyl-2-carboethoxycyclohexanone-2-β-propionate (**A**) is transformed by boiling alcoholic sodium ethoxide into ethyl 6-carboethoxycyclohexanone-2-β - propionate **(B)** [4].

The mechanism being obviously alcoholysis and ring closure in a new position as shown below:

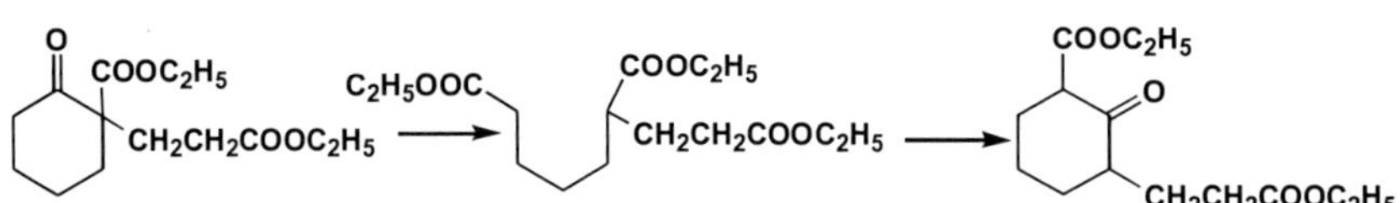

(A)

The isomeric 2-carboethoxy-2-methylcyclohexanone (**C**) is in fact, easily converted to **(D)** by sodium ethoxide [5].

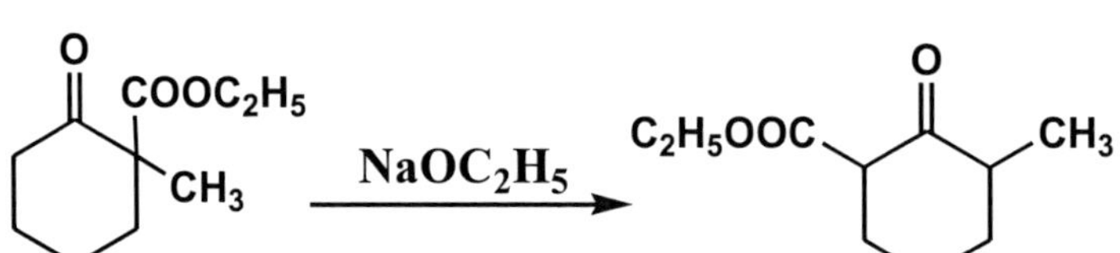

(C) **(D)**

4. Dieckmann Condensation (Intramolecular Claisen Condensation

In 1894 W. Dieckmann [6] reported that intramolecular condensation of dicarboxylic acid di-Et esters in the presence of base forms cyclic β-keto esters. The intramolecular Dieckmann condensation is a facile method for the synthesis of 5- or 6-membered cyclic compounds.

Ethyl pimelate can be cyclized to 2-carboethoxycyclohexanone [7].

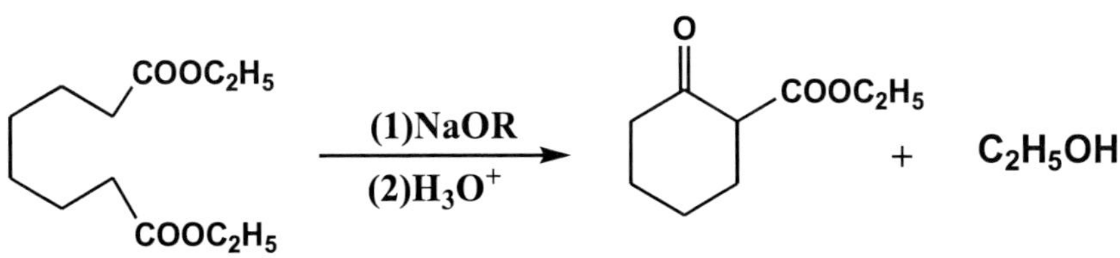

Luche et. al. [8], utilized the ultrasound in organic synthesis and postulated that in the reaction of diethylpimelate with colloidal potassium in toluene at room temperature, Dieckmann condensation is highly favoured and produce 2-carboethoxycyclohexanone.

Treatment of the diester with six or more equivelant of the organocopper reagent in THF at -20°C afforded the keto ester derivative (where; R = Me, Et, Bu, Ph, CH_2:CH) as keto-enol mixtures in yields of 80-96% [9].

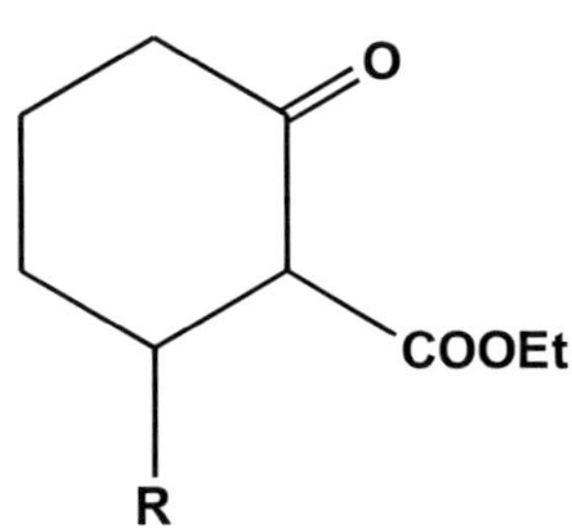

5. Miller and Nord Reaction

Miller and Nord [10] repoted the formation of (**E**) from ethyl acetoacetate and 2-thiophenecarboxldehyde.

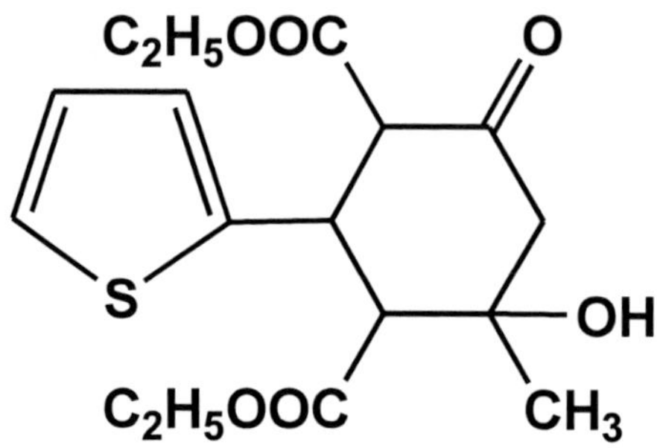

(E)

It has been reported by Lal and Bhaduri [11], that aromatic aldehydes react with methyl or ethyl acetoacetate to give 3-aryl-2,4-dicarboethoxy (or carbomethoxy) -5-hydroxy-5-methylcyclohexanones **(F)** .

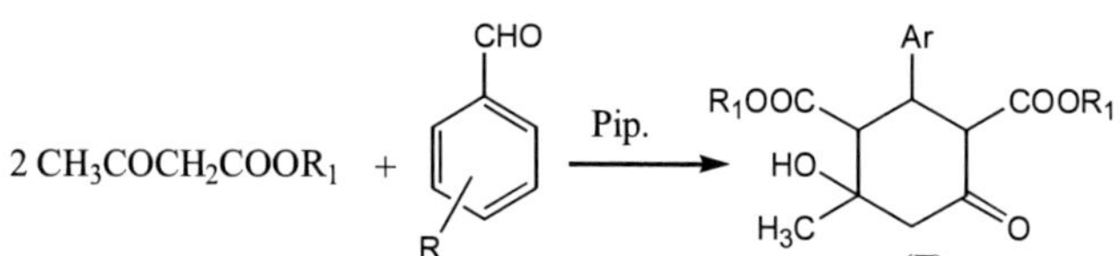

6. Esters of α,α-Diethylacetonedicarboxylic Acid React with Formaldehyde and Methylamine to Give Pyridones [12]

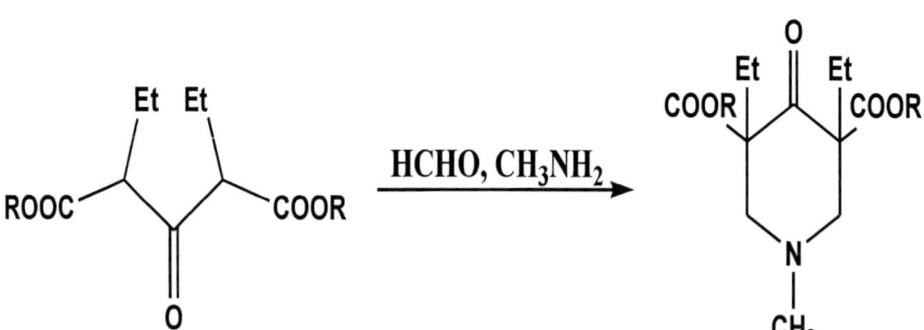

7. Cyclic β-Keto Esters of the Type (G) Can be Prepared from Cyclic Ketones and Ethyl Cyanoacetate

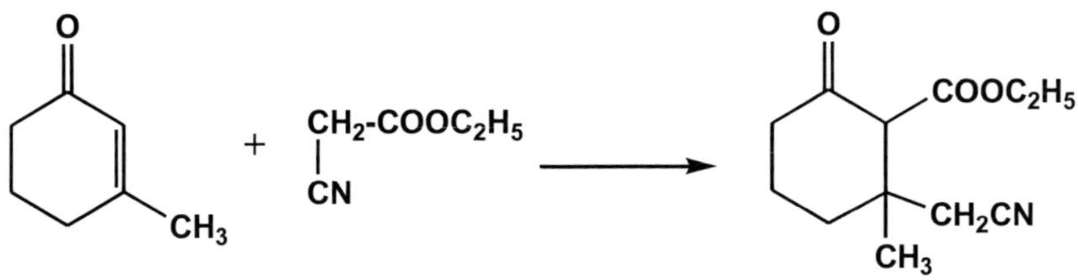

(G)

Shafer et al. [13] submitted the following alternative interpretation; the reaction between cyclohexenones and diethyl malonate gave first the product (**H**) of a Michael reaction. This compound then underwent alcoholysis (reverse Dieckmann) to give the ester (**I**) which is known to cyclize to compound (**J**).

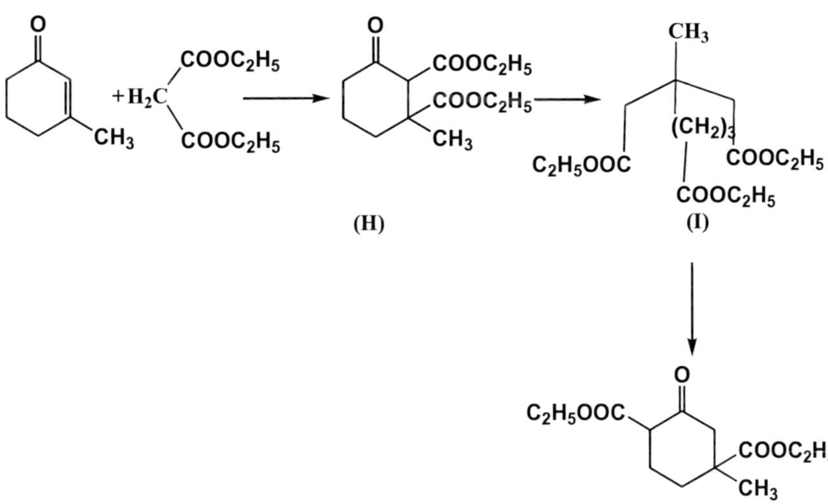

(J)

8. Claisen Acylation and Carboethoxylation of Ketones and Esters [14,15]

The use of sodium hydride [16,17] in the acylation and carboethoxylation of ketone produce good yield than the other common reagents [18].

Acylation and carboethoxylation of cyclohexanone with diethyl carbonate produce 2- carboethoxycyclohexanone [19].

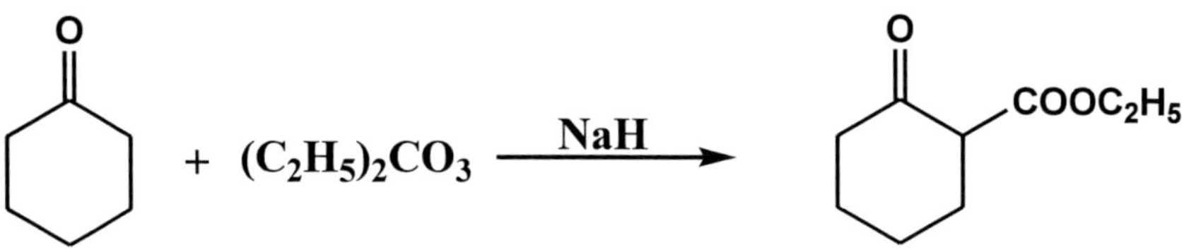

9. Homologation of Ketones with Ethyl Diazoacetate

It was mentioned that the Lewis acid catalyzed the reaction of diazoacetate with ketones to prepare some β-keto esters [20-22], in the same manner 2-carboethoxycyclohexanone was prepared from cyclopentanone and ethyl diazoacetate catalyzed by triethyloxoniumfluoroborate [23, 24].

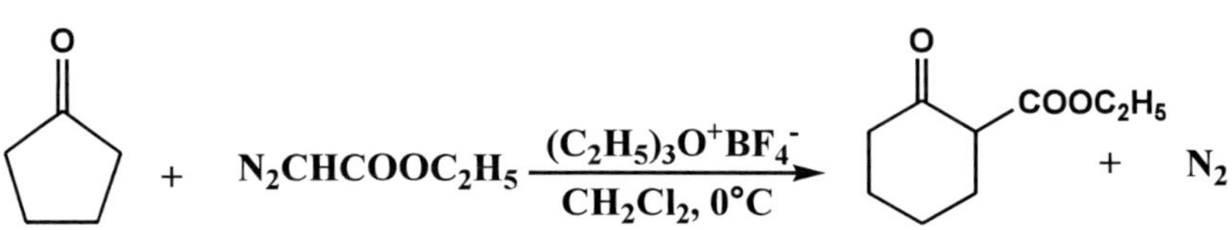

10. Synthesis of Carboethoxycyclohexanone

Shahak (1966) [25] reported the formation of 2- carboethoxycyclohexanone by prior conversion of cyclohexanone to the enolate form by the base followed by the reaction with alkyldialkoxy phosphinyl formate [26].

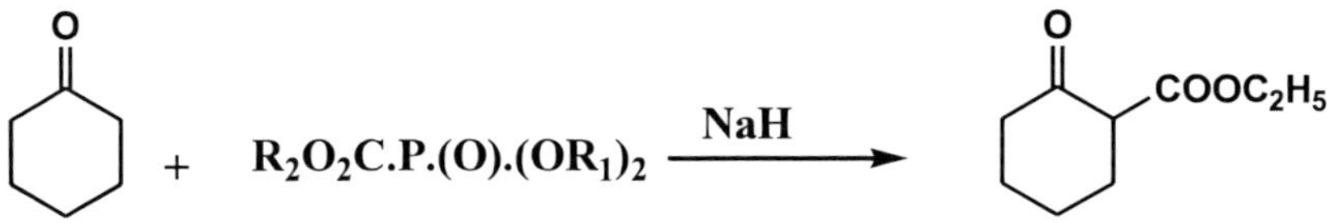

11. Synthesis of Carboethoxycyclohexanone

Kociołek et al. [27] found that the reaction of carbonylcyanide with ketone followed by alcoholysis produce 2-carboethoxycyclohexanone.

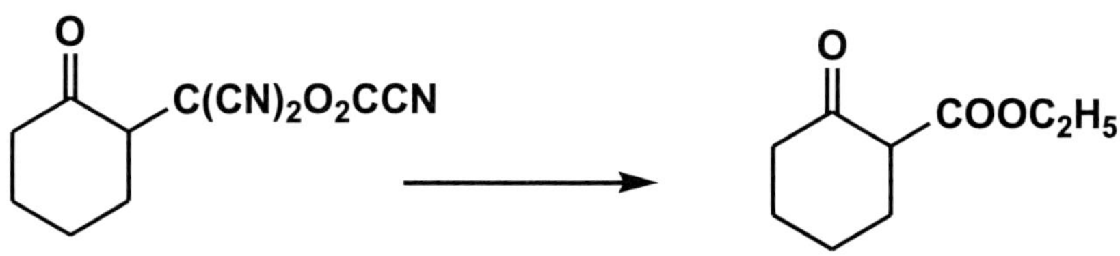

12. The Enamine Acylation and Alkylation of Carbonyl Compound

This synthetic method is remarkable by its mildness and by the ease with which monoalkylation or acylation can be achieved, and the carbon alkaylation product was hydrolysed by water to an alkylated ketone [28-30]. Thus 2-carboethoxycyclohexanone can be prepared by heating morpholine enamine [31] of cyclohexanone with ethyl chloroformate followed by stirring with 10% HCl [32,33].

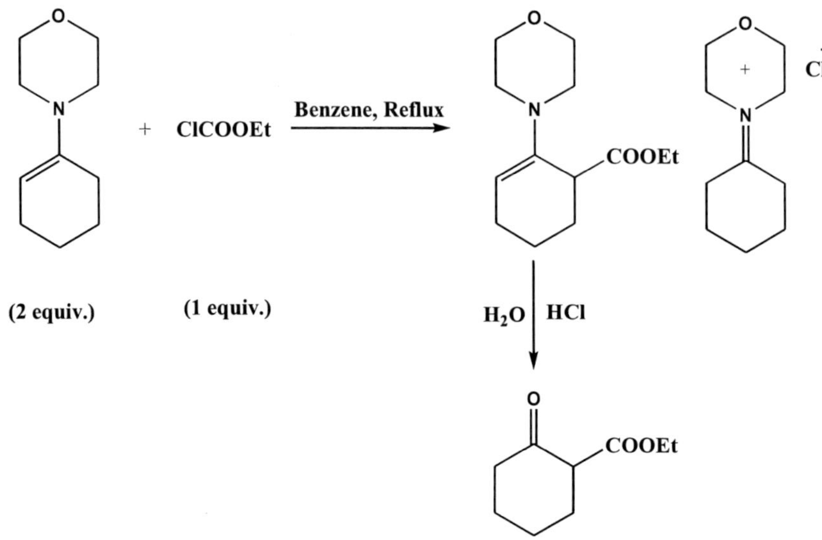

Also the hydrolysis of 1-morpholino-1-cyclohexene and diethylpyrocarbonate produce 2-carboethoxycyclohexanone [34].

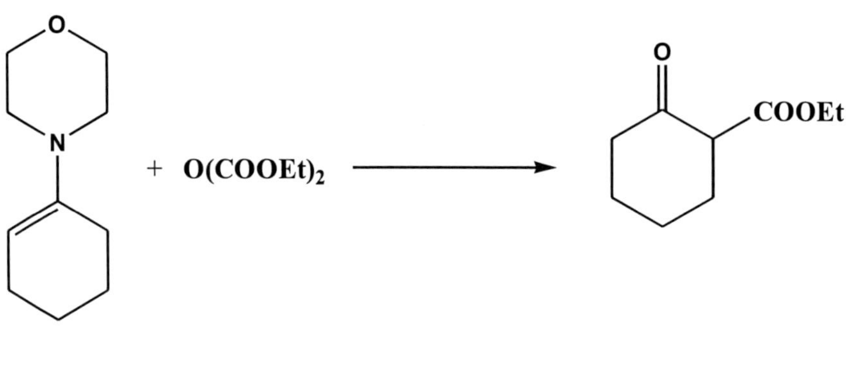

13. Synthesis of 2-Carboethoxycyclohexanone

2-Carboethoxycyclohexanone can be prepared via the reaction of cyclohexanone and diazo lithio compound [35, 36].

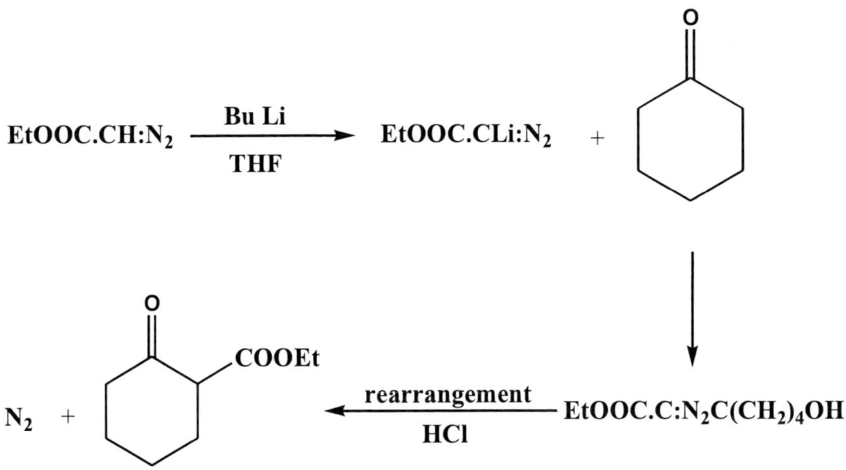

On the other hand, Hellou et.al [37] utilized diethyl dicarbonate as a convenient reagent for the prepartion of 2-carboethoxycyclohexanone.

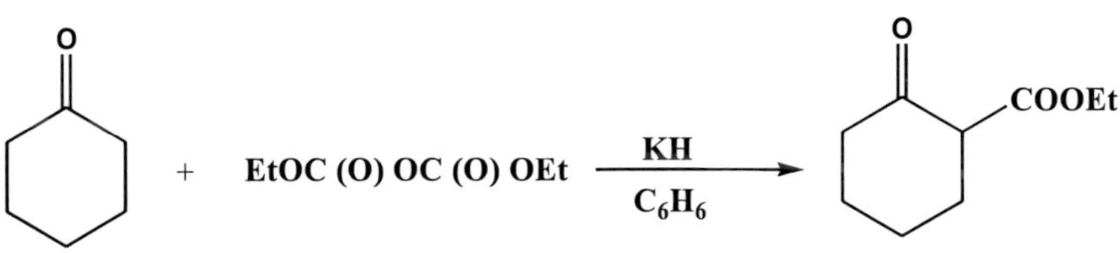

14. Synthesis of N-Containing Cyclic β-Keto Esters

Intramolecular condensation of N- containing dicarboxylic acid diesters in the presence of bases in pyridines as solvents, then salt decomposition of the resulting salts of enol form of β-keto esters by acids[38].

Refluxing 4-aminopyridine with ethyl acrylate for 4 days gave 88% diethyl 3, 3'-[N-(4-pyridyl)imino]dipropionate, which was treated with NaH in toluene and ethanol at room temperature for 1.5 h followed by refluxing with 4N HCl for 1 h to give 85% 1-(4-pyridyl)-4-piperidone [39].

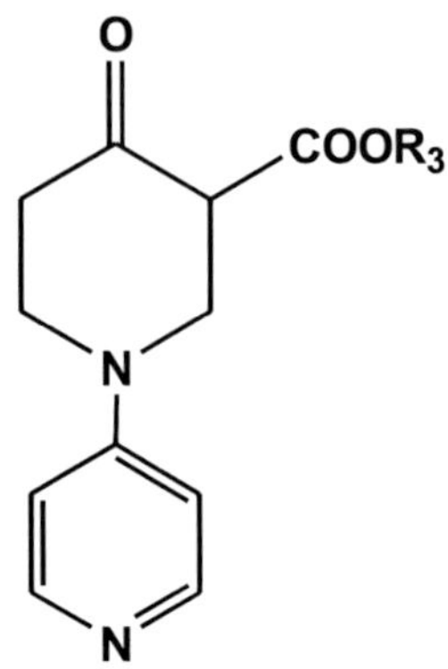

15. SYNTHESIS OF α,α-DIFLUORO-β-KETO ESTERS:

α,α-Difluoro-β-keto esters of formula $R^1COCF_2CO_2R^2$ (R^1 = alkyl, aryl, aralkyl, alkenyl, alkynyl, alkyloxy, aryloxy, alkenyloxy, alkynyloxy, aralkyloxy; R^2 = alkyl, aryl, aralkyl, alkenyl, alkynyl) are prepared by coupling reaction of acid halides of formula R^1COX^1 (R^1 = same as above; X^1 = halo) with difluoroketene silyl acetals of formula $F_2C{:}C(OR^2)OSiR^3R^4R^5$ (R^2 = same as above; R^3, R^4, R^5 = alkyl which together optionally form a cyclic group, aryl, aralkyl, alkenyl, alkynyl, alkoxy, aryloxy, alkenyloxy, alkynyloxy, aralkyloxy) in the absence of a catalyst or in the presence of a metal catalyst [40].

Chapter 2

REACTIONS

1. ALKYLATION

β-Ketoesters are easily alkylated [41], ethylcyclohexanone-2-carboxylate **(1)** [42] reacts with alkylhalide **(2)**. The reaction product **(3)** undergoes ester elimination to give the 2-susbtituted cyclohexanone **(4)**, that can be cyclized to a β-decalenone **(5)**. The product **(3)** is cyclized also with sulphuric acid to afford (**6**).

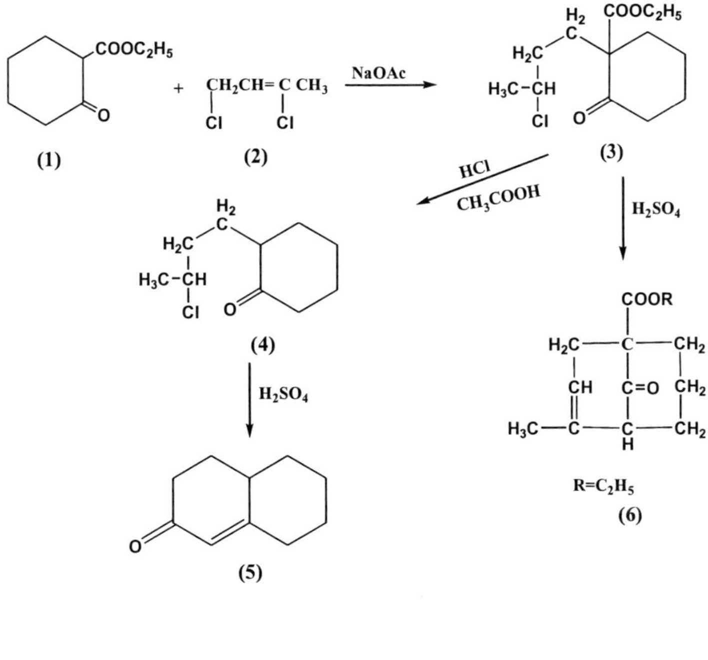

(1) (2) (3) (4) (5) (6)

Diethyl 1,4-cyclohexanedione-2,5-dicrboxylate **7** is very well adapted to the preparation of the bicyclo system; diethyl bicyclo [1.2.2] heptane **(8)** , diethyl bicyclo [2.2.2]octane **(9)** and diethyl biethyl bicyclo [3.2.2] nonane **(10)** dionedicarboxylate, by treating its disodio derivative with methylene iodide, 1,2-di-bromoethane, 1,3-dibromopropane respectively [43].

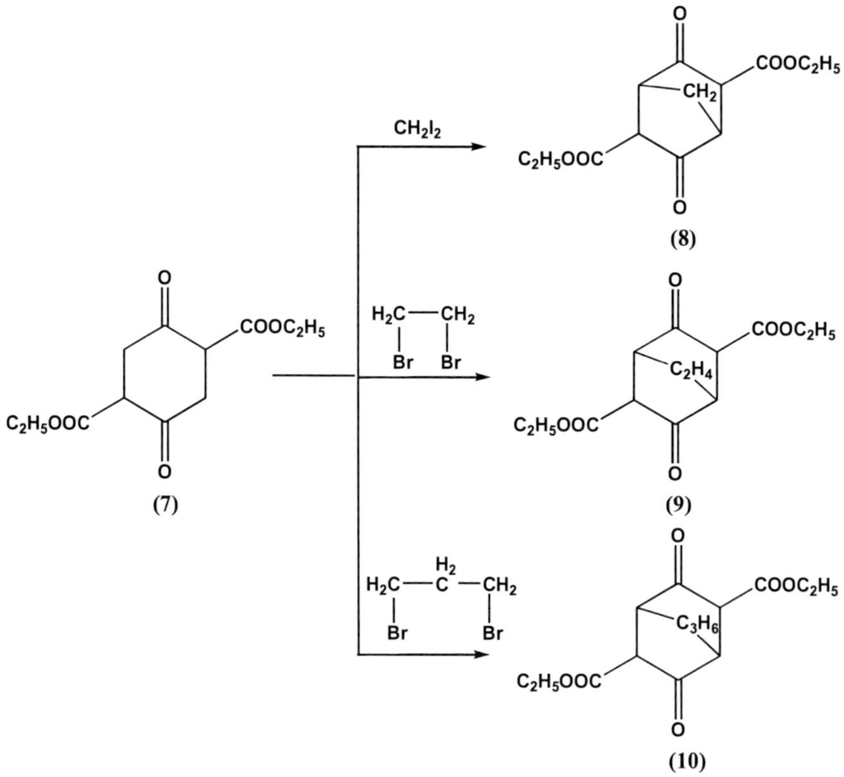

An interesting alkylation of **(7)** with methyl iodide and/or ethyl iodide gave diethyl 2,5-dimethylcyclohexane-1,4-dione 2,5-dicarboxylate **(11a)** and 3,5-diethylcyclohexane 1,4-dione- 2,5dicarboxylate **(11b)**. A mixture of the two isomers is obtained in both cases **(11a)** and **(11b)**. The ester groups propably being cis or trans with respect to the cyclohexane ring. After saponification and decarboxylation, no signs of isomers are observed **(12a,b)** [44].

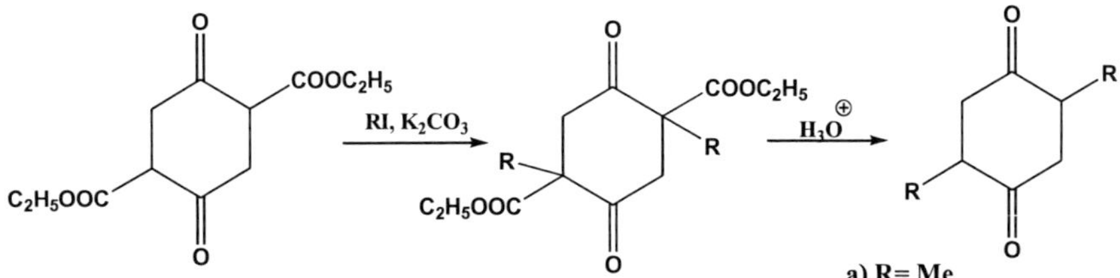

(7) (11a,b) (12a,b)

A new C,O-cycloalkylation of dilithiated cyclic β-keto esters and β-keto sulfones with 1,4-dibromo-2-butene is reported which results in regio- and diastereoselective formation of 2-vinyl-2,3,3a,4,5,6-hexahydro-2,3-benzofurans. The products could be efficiently transformed into functionalized bicyclo[3.2.1]octan-8-ones by a palladium-catalyzed rearrangement reaction [45].

Treatment of oxocyclohexanecarboxylate enamines **(13)** and **(14)** with H_2C:CHCHO in refluxing dioxane gave the (formylethyl)cyclohexanone **(15a)** and its cyclic tautomer **(15b)** in 51% yield [46].

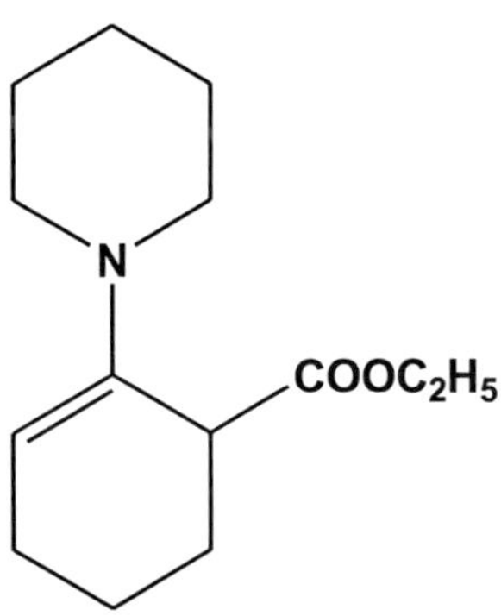

(13)

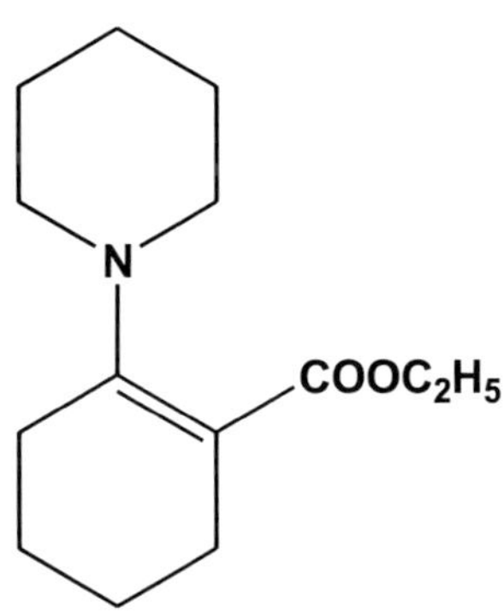

(14)

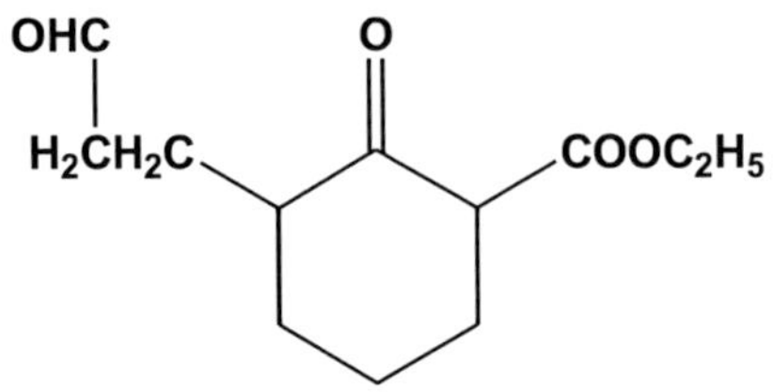

(15a)

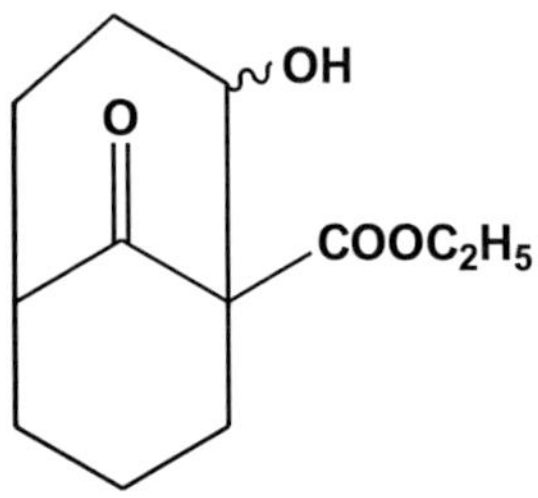

(15b)

The open chain analog, ethyl α-acetylvalerate undergoes C-alkylation exclusively with β-dimethylaminoethyl chloride to give Et α-(β-dimethylaminoethyl)-α-acetylvalerate [47].

Treatment of cyclic β-keto esters with alkyl halide under mild reaction conditions afforded α-alkylated β-keto esters[48] **(16)**.

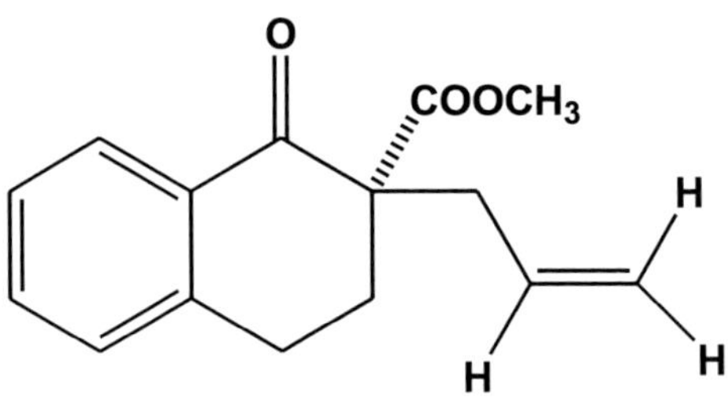

(16)

Reactions of various cyclic β-keto esters and 1,3-diketones with alkynes in the presence of $Pd_2dba_3.CHCl_3$ (5 mol %), PPh_3 (40 mol %), and PhCOOH (10 mol %) proceeded at 100° C in toluene (5 M) to give the corresponding allylation products in high yields in a regio- and stereoselective manner [49].

An efficient and mild acyl-alkylation of arynes generated from o-silyl aryltriflates (**17**; $R^1 = R^2 = R^3$ = H; R^1 = MeO, $R^2 = R^3$ = H; $R^1 = R^2$ = H, R^3 = Me; R^1 = H, $R^2 = R^3$ = OCH_2O) with β-keto esters $R^4COCH_2CO_2R^5$ (R^4 = Me, Et, Me_2CH, Ph, $PhCH_2$, $PhCH_2OCH_2$; R^5 = Me, Et, menthyl, cholesteryl) gave alkoxycarbonylmethyl-substituted aromatic ketones **(18)** in 75-99% yields [50].

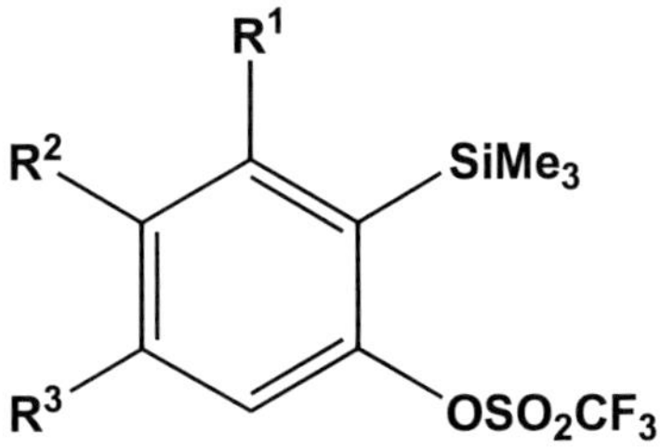

(17)

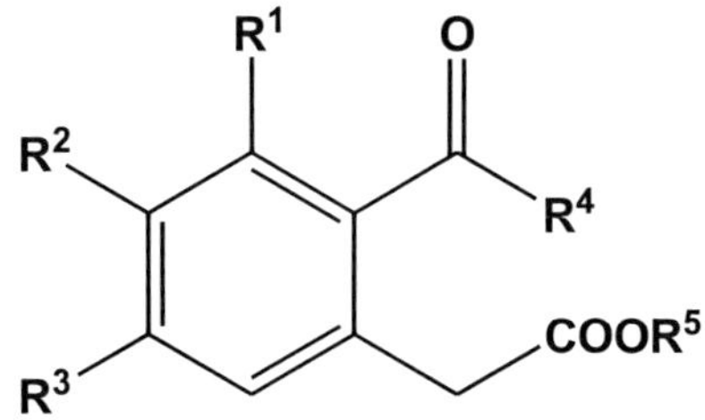

(18)

The substitution of an α,α-difluoro-β,β-dichloroethoxy group for methoxy in the Torgov steroid intermediate 6-methoxy-1-vinyl-1-tetralol, abolishes the tendency of the tertiary alcohol (which is also allylic and benzylic) to ionize during acetylation, where the resulting vinylic tertiary acetate obtained can easily be converted to the corresponding π-allyl palladium acetate complex. This reagent can then be used to alkylate the anions of cyclic β-keto-esters in high yield [51].

Alkylation of cyclic β-keto esters **(19** and **20;** R = H) with optically active sulfonium salts, e.g., $R^1S+MeEt$ X- (R^1 = Ph, 4-ClC_6H_4; X- = ClO_4-, d-camphorosulfonate) gave products **(19** and **20**; R = Me, Et), respectively, with high stereoselectivity [52].

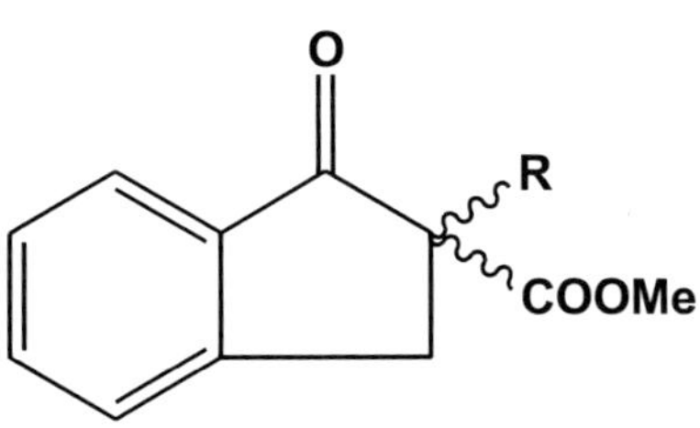

(19)

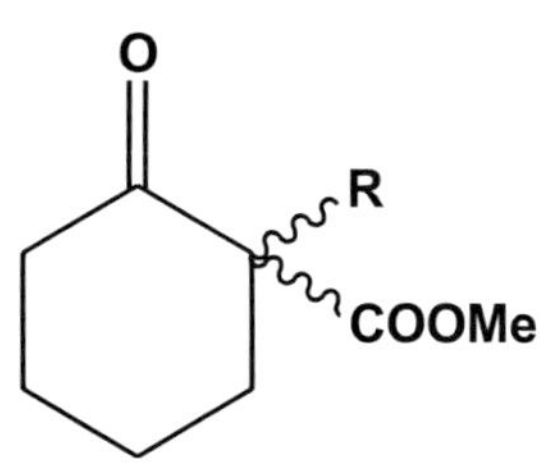

(20)

Alkylation of cyclic β-keto ester **(21)** with racemic sulfonium salts containing optically active alkyl groups, e.g., (S)-2-octyl and (S)-CHMeEt, gave C-alkylated products with inversion of configuration at the asymmytric alkyl C [53].

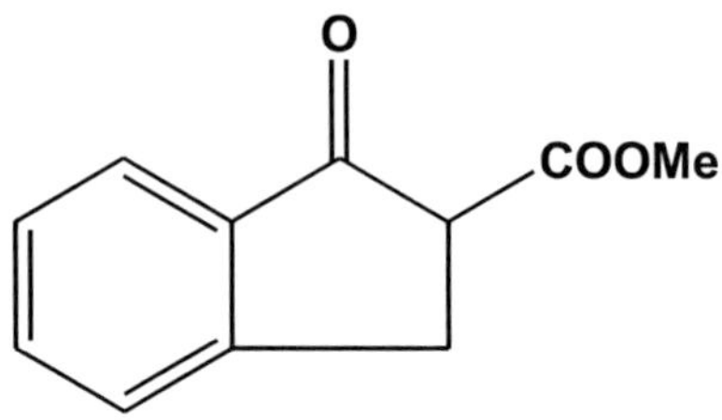

(21)

2. MICHAEL REACTION

An alternative method of alkylation, is Michael condensation in high β-keto ester can act as a Michael donor (keto form) as a Michael acceptor (enol form).

It has been reported that **(1)** was condensed with phenylvinyl sulfoxide in sodium hydride/THF and in toluene to give **(22)** and **(23)** respectively[54].

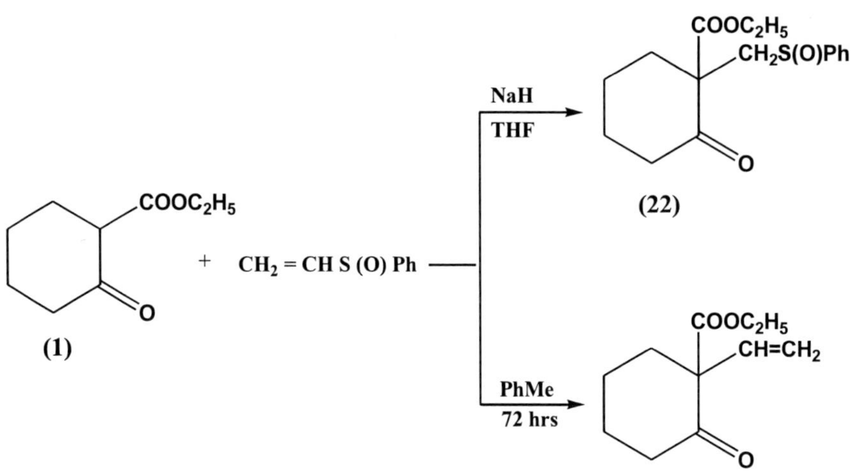

Acrylonitrile when used as an acceptor [55, 56] in the Michael reaction with ethyl cyclohexanone-2-carboxylate **(1)**, using a catalyst gave **(24)** in 85% yield.

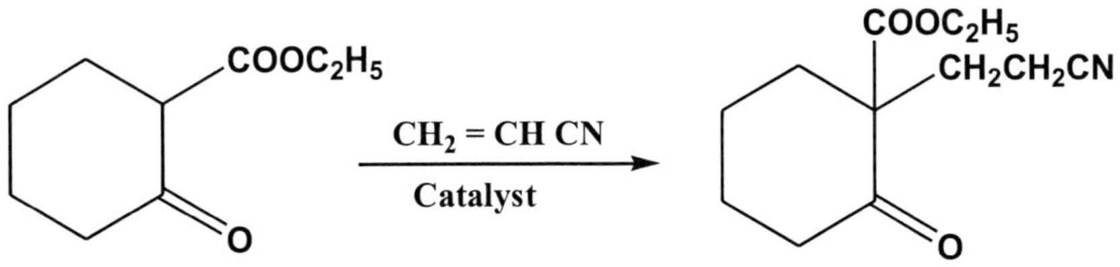

(1) (24)

Ethyl potassiocyclohexanone-2-carboxylate **(1)** condenses with ethylideneacetone **(25)** in dry alcoholic solution on cold to yield **(26)** (R=CH_3) [57].

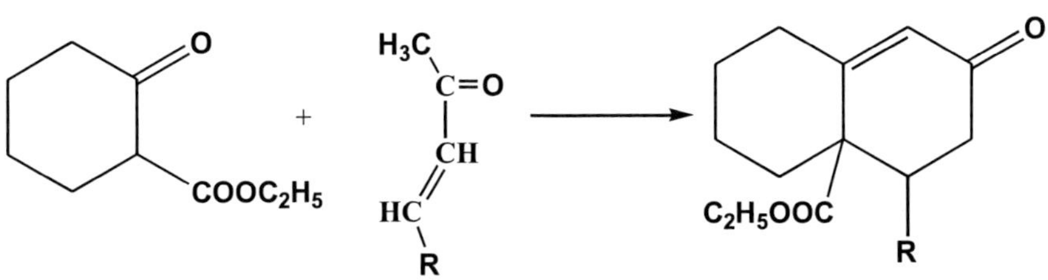

(1) (25) (26)

Ethyl cyclohexanone-2-carboxylate **(1)** adds to α,β-ethylenic aldehydes like acroline at - 70° to give **(27)** [57]. This reaction yielded only polymeric prducts under usual conditions of Michael reaction. When the aldehyde **(27)** was treated with acids, compound **(28)** was formed. Both **(27)** and **(28)** yielded 1-carboethoxybicyclo [3.3.1]- non-3-en-9-one **(29)** on treatment with concentrated sulphuric acid [58].

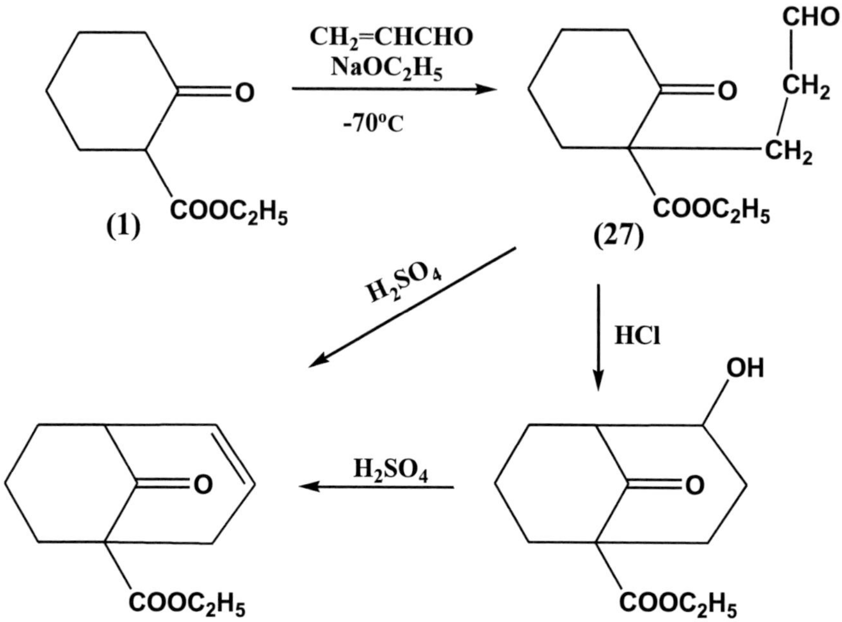

(29) (28)

Dreiding et al. [59] reacted ethyl cyclohexanone-2-carboxylate **1** with ethylvinyl ketone in the presence of benzyltrimethylammonium hydroxide, and obtained (**30**) in a good yield.

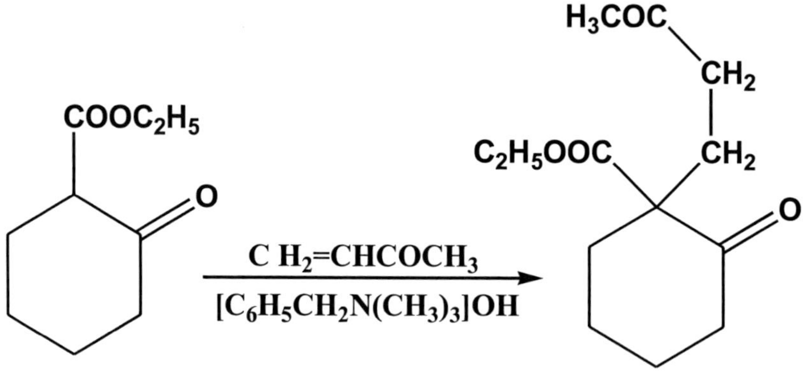

(1) (30)

Substituted cyclohexanonecarboxylate, like 2-carboethoxy-3,3-dimethylcyclohexanone **(31)** reacts with ethylacrylate in the presence of sodium ethoxide to give **(32)** [60].

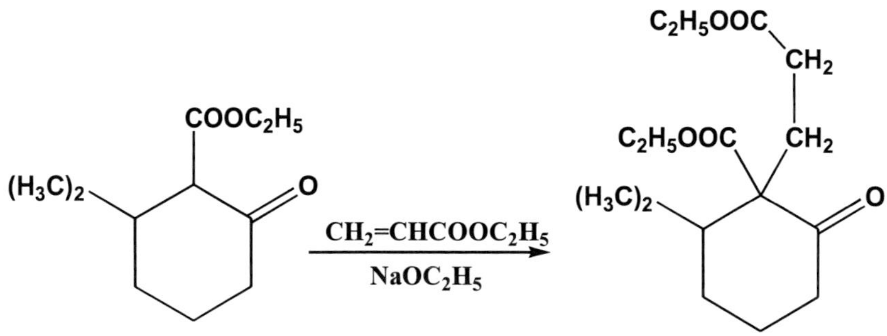

(31) **(32)**

Michael condensation of **(7)** with equivalents of benzalacetone afforded diethyl 2, 3, 4, 6, 7, 8, 9, 10-octahydro-2,6-dioxo-4,8-diphenyl-4a, 8a-anthracene-dicarboxylate **(34)** via a secondary condensation between the carbonyl group of the original donor and the terminal methyl group of the initial Michael addut **(33)** which was not isolated [61].

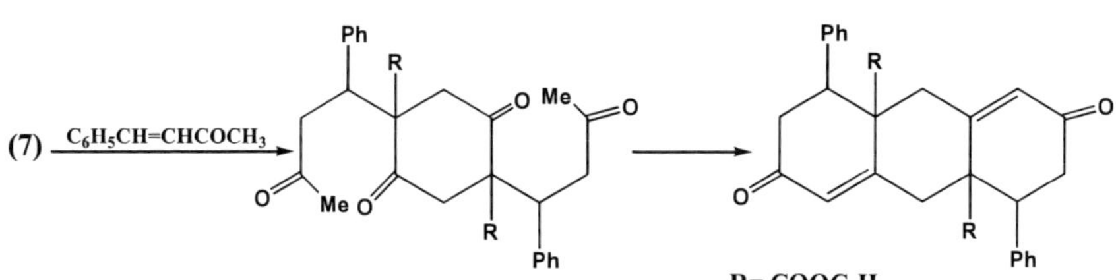

(33) **(34)**

On the other hand, diethyl 1,4-bis (substituted-α-phenacylbenzyl)-2,5-dioxo-1,4-cyclohexane dicarboxylate **(35)** were obtained by condensing **7** with a variety of α,β-unsaturated ketones [61].

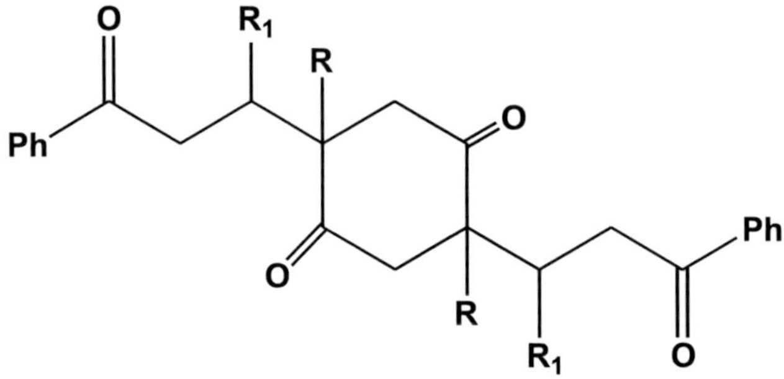

(35)

Cyanoethylation of **(7)** gave a single product which was identified as 2,3,3, 5, 6, 6-hexacyanoethy- diethyl-1,4-cyclohexanedione-2,5-dicarboxylate **(36)**.

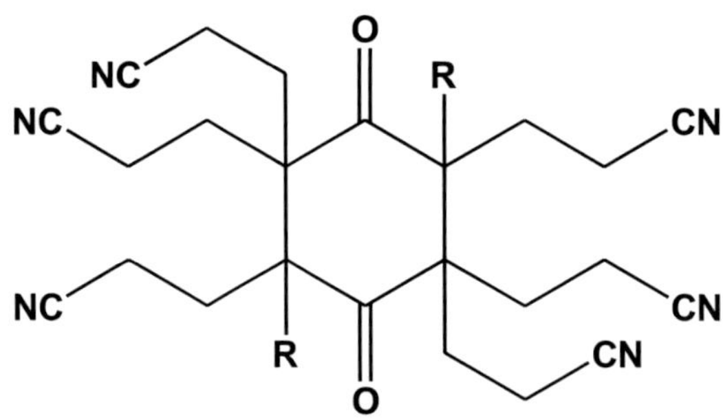

(36)

Metwally et al. [41] reported that when 3-phenyl-2,4-dicarboethoxycyclohexanone **(37)** was treated with anisylcyclohexanone in ethanolic sodium ethoxide, the Michael adduct **(38)** was obtained.

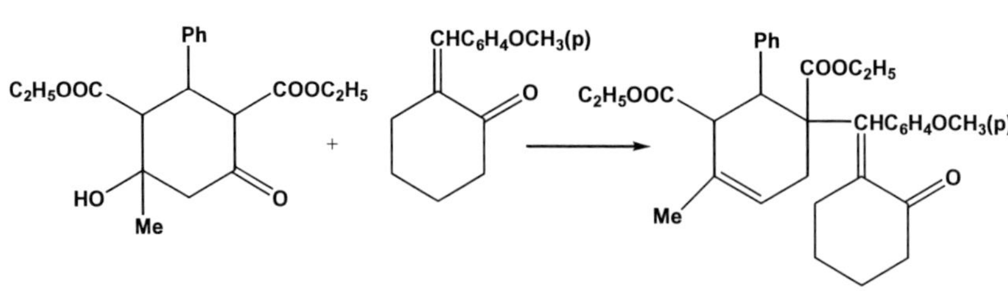

(37) (38)

A catalytic enantio- and diastereoselective Michael reaction was achieved to construct vicinal quaternary and tertiary carbon centers in one step. Using 5 mol % of La(O-i-Pr)$_3$ and 10 mol % of a new N-linked-BINOL type ligand, the reaction of α-substituted β-keto esters to cyclic enones provided the corresponding Michael adducts in up to quantative yield with a diastereomeric ratio up to 86/14 and enantiomeric excess up to 86% for the major isomer [62].

Enantioselective Michael additions of cyclic β-ketoesters to Me vinyl ketone catalyzed by cinchona alkaloids revealed that the induced enantioselectivity was significantly influenced by both the structure of the catalyst and that of the substrate. High enantioselectivities were obtained in the reaction of ethyl 2-oxocyclopentanecarboxylate and Et 2-oxocyclohexanecarboxylate (up to 83 and 80%, respectively) at a low cinchona :reactant ratio of 1:500 [63].

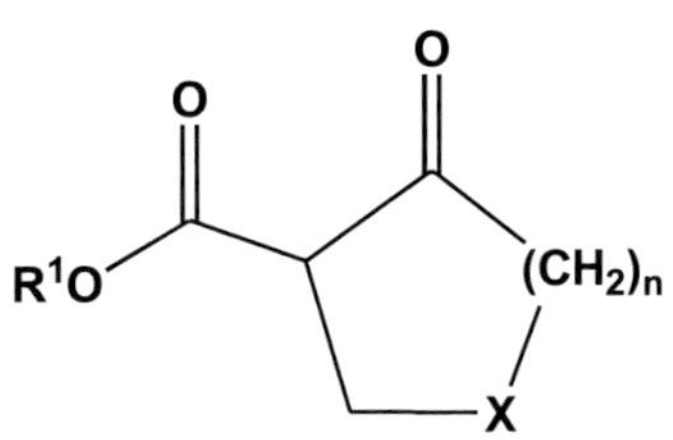

(39)

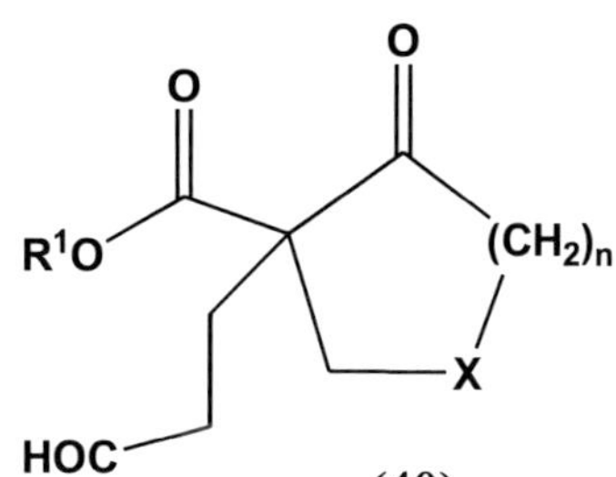

(40)

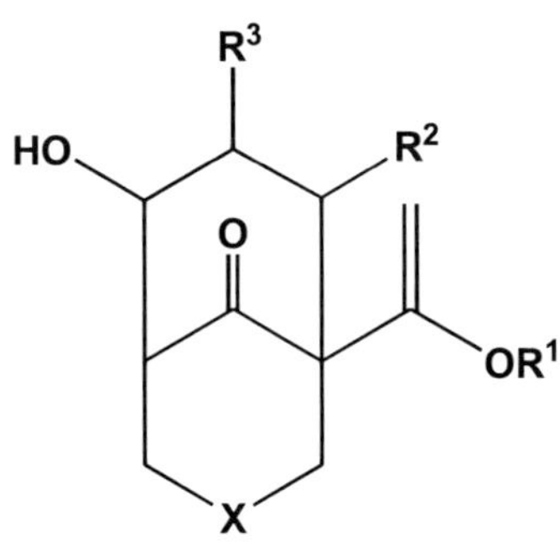

(41)

Depending on the nature of the Dowex basic ion-exchange resin, cyclic β-keto esters (**39**; X = CH_2, $PhCH_2N$; R^1 = Me, Et; n = 1, 2) react with α,β-unsaturated aldehydes $R^2CH{:}CR^3CHO$ (R^2, R^3 = H, Me) to give either the corresponding Michael adducts **(40)** or highly functionalized bicyclo[3.3.1]nonanes **(41)** [from (**39**; n = 2)] in a one pot Michael addition - intramolecular aldolization sequence [64, 65].

Quaternary ammonium hydroxides were immobilized on a polystyrene-poly(ethylene glycol) copolymer resin. The amphiphilic polymer-supported ammonium hydroxides catalyzed Michael reaction of cyclic β-keto esters with several Michael acceptors in water to give corresponding adducts in high yields [66].

The asymmetric copper(II)-catalyzed Michael reactions of cyclic β-oxo esters (Et cyclopentanecarboxylate and Et cyclohexanecarboxylate) with Me vinyl ketone, yielding the corresponding addition products **42** (n = 0, 1) with a quaternary stereocenter in selectivities up to 99% ee, which undergo simple precepitation from diethyl ether to yield novel polymer-bound L-valine derived auxiliary **(43)** [67].

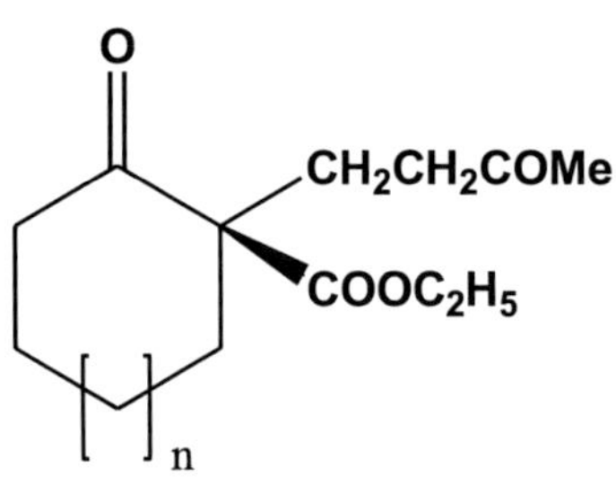

(42)

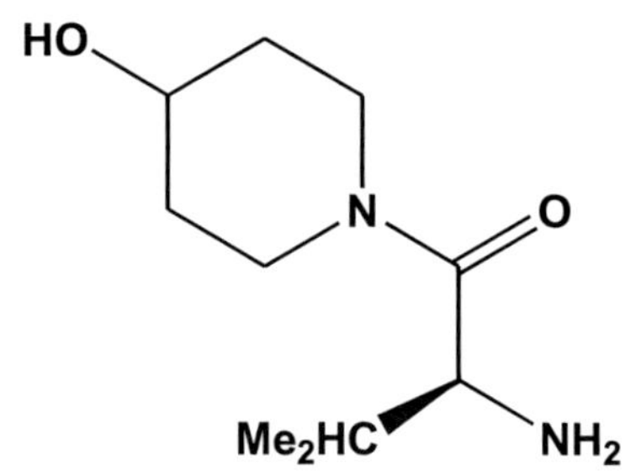

(43)

The newly developed La complexes with (S, S)-NR-linked-BINOLs (X = NH, NMe) afforded the best results in terms of reactivity and selectivity as a general catalytic asymmytric Michael addition of acyclic β-keto esters to cyclic enones, in which asymmytric induction occurs at the β-position of the acceptors. In general, **(44a)** (X = NMe) was suitable for the combination of small enones and small β-keto esters, and (X = NH) was suitable for bulkier substrates (steric tuning of the catalyst) [68].

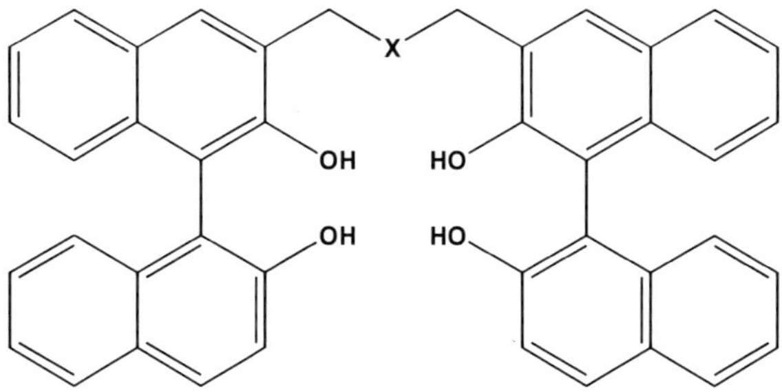

44a

Michael addition products prepared from 1,3-diketones or β-keto esters and α,β-unsaturated ketones are irradiated with microwaves on supports in the presence of bases to give polycyclic unsaturated ketones, useful for pharmaceuticals, agrochemicals, or their intermediates [69].

The bis[[(1R,1'R)-2,2'-dihydroxy-1,1'-binaphthyl-3-yl]methyl] ether-lanthanum complexes **(44b)** (M = H, monovalent metal atom) catalyze asymmytric Michael addition reaction of β-dicarbonyl compounds with cyclic or noncyclic enones. They are very stable even in air and can be stored for a long period of time and reused [70].

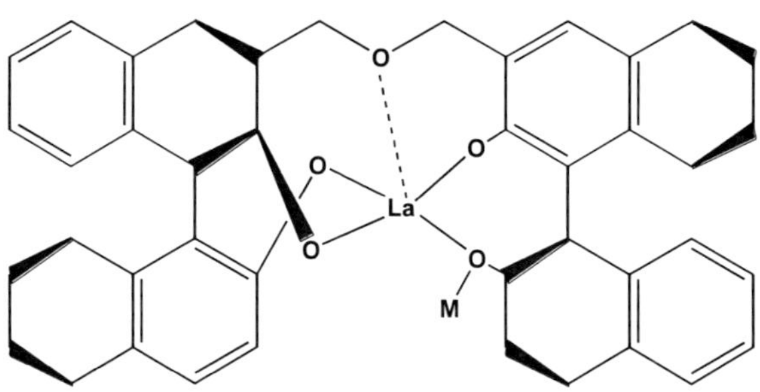

44b

Seven-membered ring annulation, e.g., **(46)** from **(45)** is achieved by an intramolecular Michael reaction catalyzed by iron (III) chloride hexahydrate. This method fails for a 17-membered ring annulation, which could be achieved if moisture is excluded and anhydrous $FeCl_3$ and methanol-free CH_2Cl_2 are used [71].

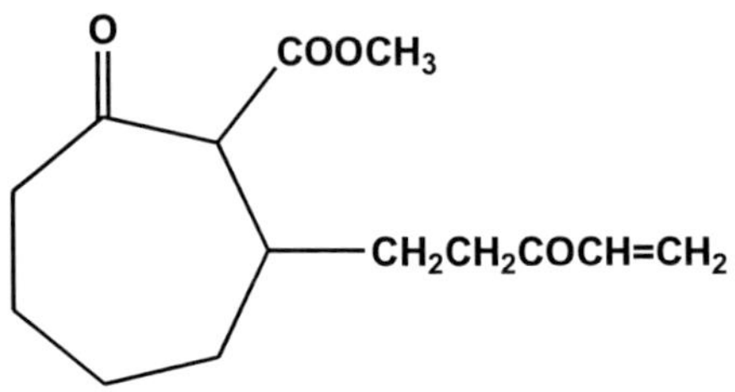

(45)

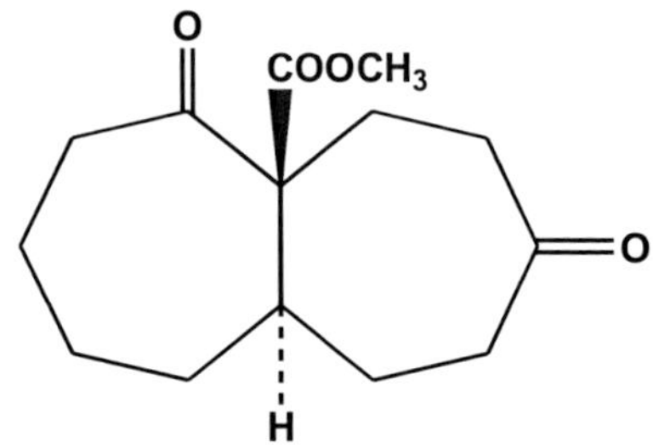

(46)

The enolates of cyclic β-keto esters react with electrophilic acetylenes to give the corresponding Michael adducts in good yields when the reaction is performed in acetone in the presence of catalytic amounts of K_2CO_3. The Michael adducts resulting from ethynyl methyl ketone, when refluxed in toluene in the presence of catalytic amounts of TsOH, undergo an intramolecular Aldol reaction leading mainly to bicyclo[n.3]alkadienones besides Robinson annulation products [72].

The intramolecular Michael addtion reaction of oxo ester ynones [**47**; m = 1, 2; n = 0-3; R = H, Me, $(CH_2)_3OAc$] gave products such as **(48)** [73].

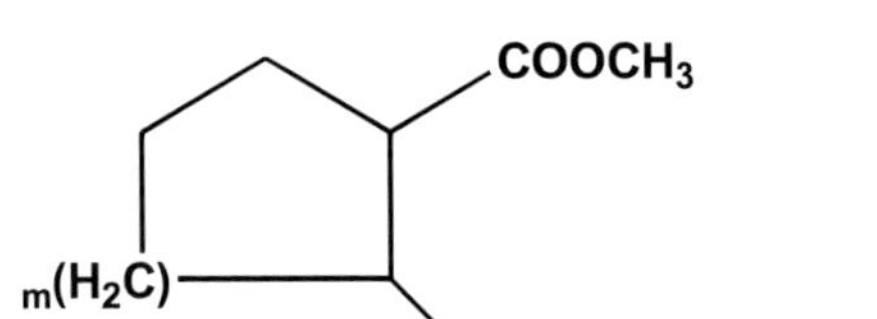

(47)

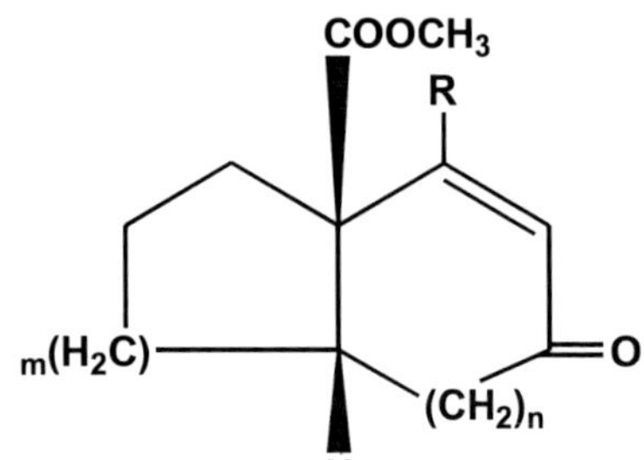

(48)

A study on the base-catalyzed intramolecular Michael addition of cyclic β-ketoester enones (**49**and **50**; n = 0-3, R = R^1 = H and/or CH_3**)** is reported [74].

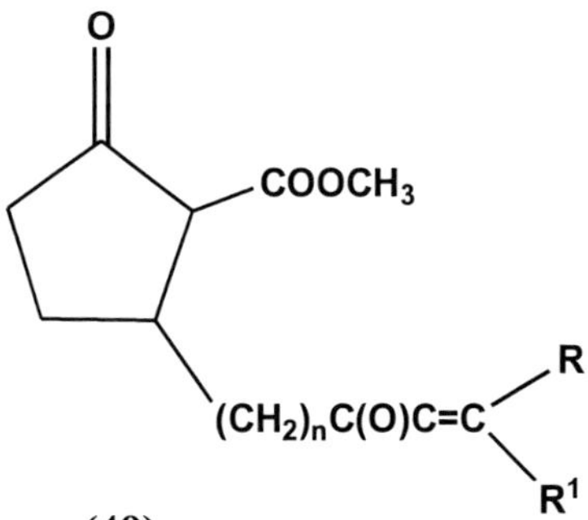

(49)

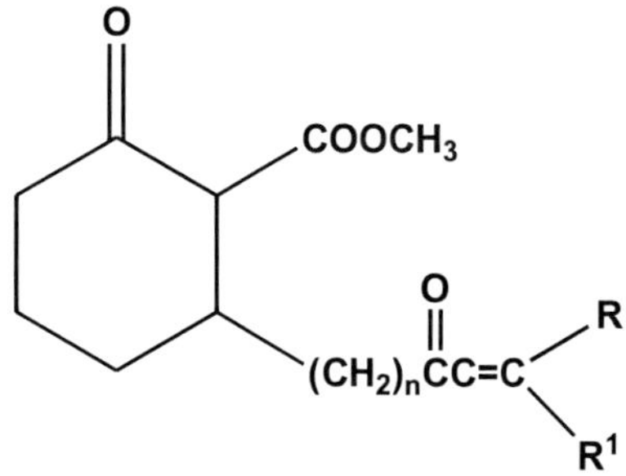

(50)

Michael alkylation of 2-cycloalkenones with R^2CuLi (R = Me, Bu, $PhCH_2$, CH_2:CH) and treatment with $ClCO_2Me$ gave the enol carbonates of cyclic β-keto esters (and in some cases the O-acylation products) which, with NaOH or NaOMe, underwent retro-Dieckmann cleavage to β-alkylalkanedioic acids or di-Me esters [75].

A facile base promoted tandem Michael addition-retro Dieckmann reaction leading to polyfunctionalized acyclic derivatives is observed by condensation of β-keto esters with α,β-unsaturated electrophiles in methanol [76].

Ru amido complexes effected asymmetric Michael addition of β-keto esters to 2-cyclopenten-1-one to give the corresponding Michael adducts [77].

3. Acylation and Arylation

When the metal salt of **(1)** was allowed to react with halides, it is possible to get the C and O-acylation products **(51)** and **(52)** respectively.

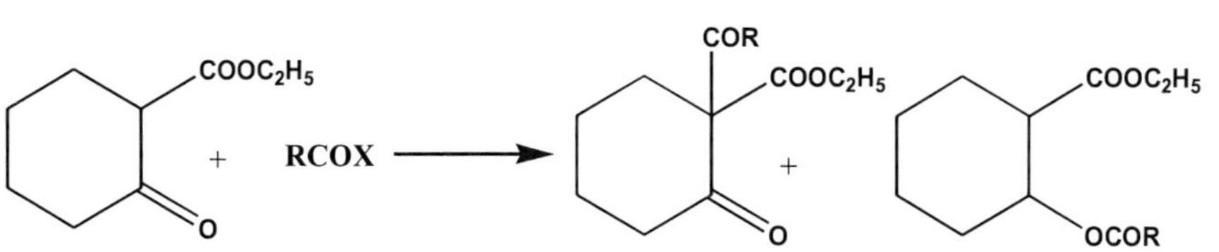

(1) (51) (52)

The amount of C and O-acylation was determined when the magnesium chelate of **(1)** [78, 79] was allowed to react in THF instead of ether with ethyl chloroformate [80], chloroacetyl chloride, acetyl chloride, acetyl bromide, acetyl perchlorate and propionyl chloride. It was found that the first two acid halides gave C-acylation.

Similary, **(1)** could be arylated with 2-bromobenzoic acid in presence of NaH and cuprous bromide as a catalyst to give **(53)** which underwent basic cleavage to **(54)** [81].

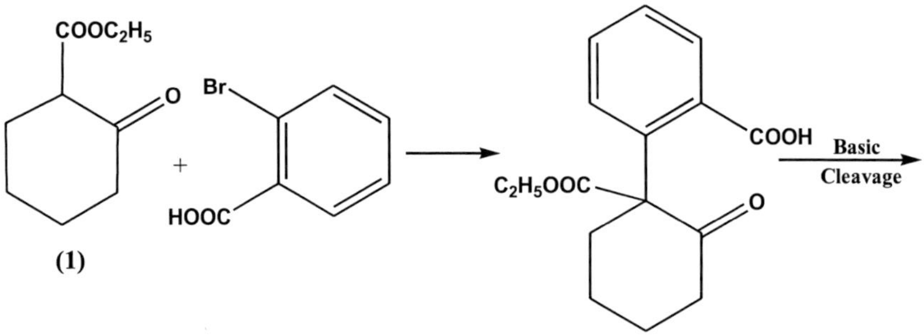

(1) **(53)**

o-$H_2OC.C_6H_{4.}CH(COOH)\ (CH_2)_4COOH$

(54)

Intramolecular acylation of an α-sulfinyl carbanion intermediate of cyclic β-keto esters **(55;** R = Me, Et; n = 1, 2, 4**)** gave spiro[4.n]alk-2-ene-1,6-diones and spiro[5.n]alk-2-ene-1,7-diones **(56** and **57**;n=1,2,4**)** [82].

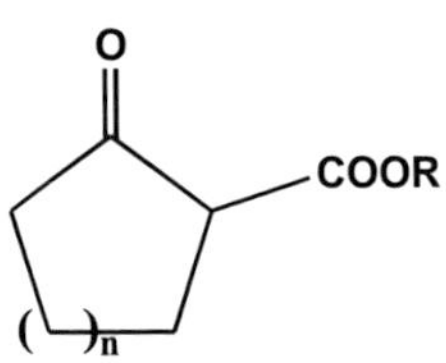

(55)

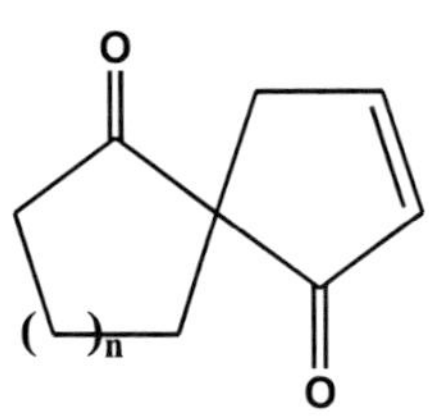

(56)

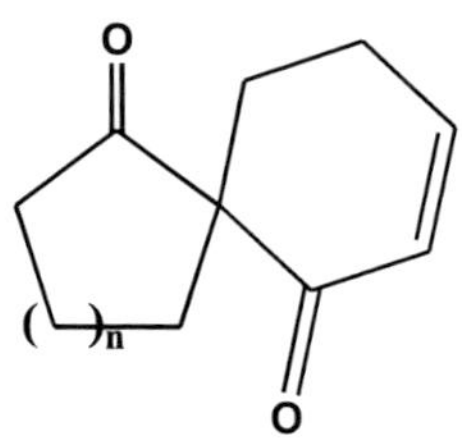

(57)

4. MANNICH REACTION

The alicylic ketones are known to undergo a double Mannich condensation. In such cases the methylenes on both sides of the carbonyl groups are aminomethylated [83, 84]. Arya [85], utilized this reaction for the synthesis of 3-azabicyclo [3.3.1]-nonane **(59)** from ethyl 2-methylcyclohexan-1-one-6-carboxylate **(58)**.

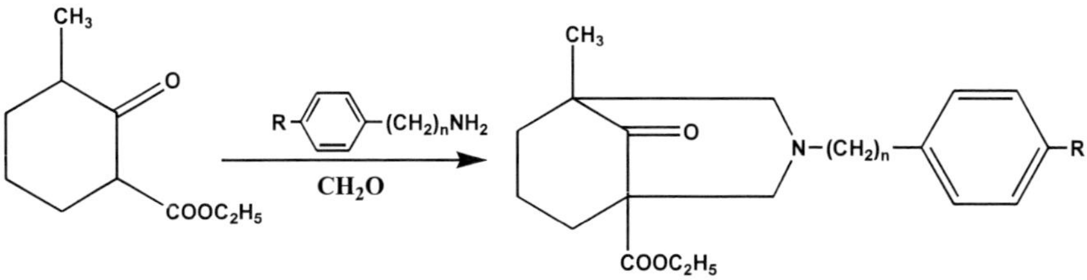

(58) **(59)** **(R=HORF n= 1-3)**

Metwally et al. [86] reported the formation of diethyl-3-azabicyclo[3.31] nonane derivatives oxalate salts **(60)** and **(61)** through the bouble Mannich reaction of **(37)** with formaline and benzylamine or cyclohexylamine.

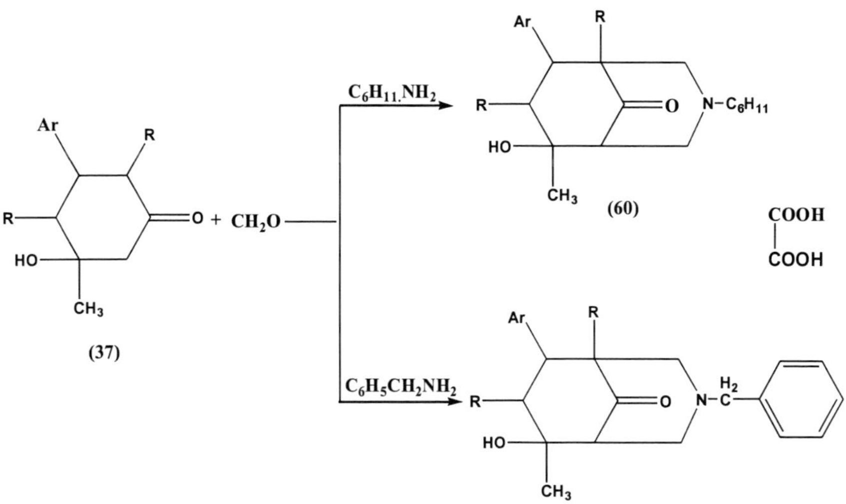

(37) **(60)** **(61)**

R= COOC$_2$H$_5$

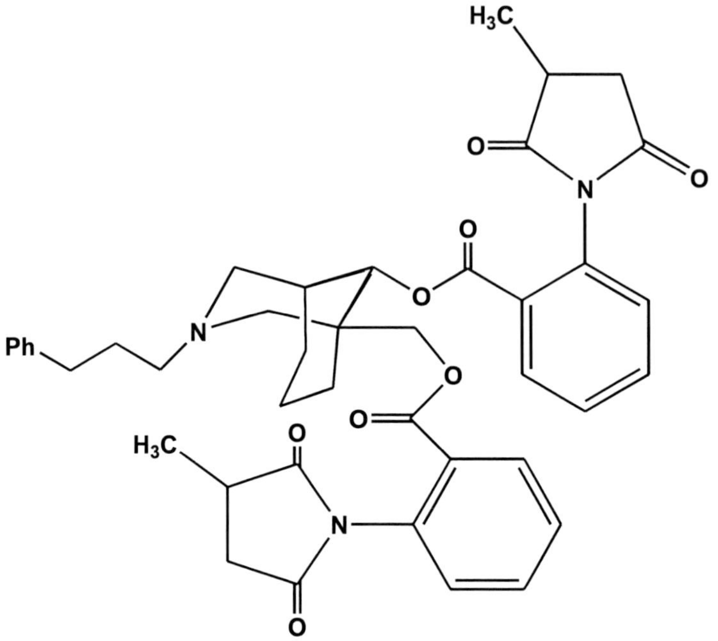

(62)

The double Mannich reaction of cyclic β-keto esters with bis(alkoxymethyl)alkylamines provides an efficient and versatile method for the construction of azabicyclo[3.3.1]nonanes and azabicyclo[3.2.1]octanes. The optimum conditions for efficient reaction involve use of the activator trichloromethylsilane in acetonitrile as solvent at ambient temperature The utility of this synthetic method is further demonstrated by the facile synthesis of several AE ring analogs **62** and **(63)** of the alkaloid methyllycaconitine by appendage of the key N-(methylsuccininimido)anthranilate pharmacophore to the N-(3-phenylpropyl)-substituted double Mannich adducts **(64)** and **(65)** [87].

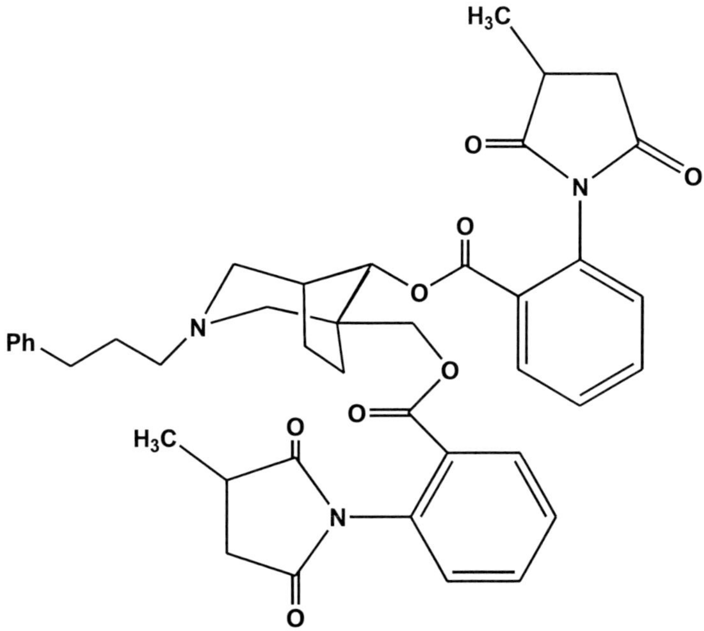

(63)

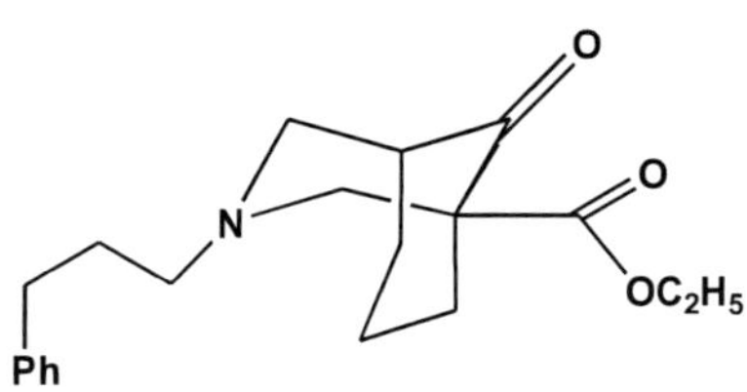

(64)

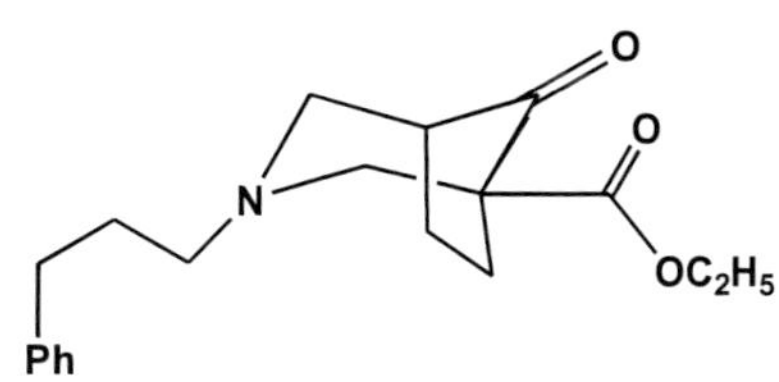

(65)

The novel pyrrolidine-sulfonamide, trifluoro-N-[[(2S)-2-pyrrolidinyl]methyl]methanesulfonamide has been prepared and used successfully to catalyze Mannich-type reactions between ketones and α-imino esters. For example, the stereoselective Mannich reaction of 1,4-dioxaspiro[4.5]decan-8-one with [[(4-methoxyphenyl)methyl]imino]acetic acid ethylester (α-imino acetate) gave (7S)-α-[(αS)-[(4-methoxyphenyl)methyl]amino]-1,4-dioxaspiro[4.5]decane-7-acetic acid ethylester **(67)** in 78% yield and in 96% enantiomeric excess. The aminooxylation of cyclohexanone with (nitroso)benzene catalyzed by **(66)** gave (2R)-2-[(phenylamino)oxy]cyclohexanone [88].

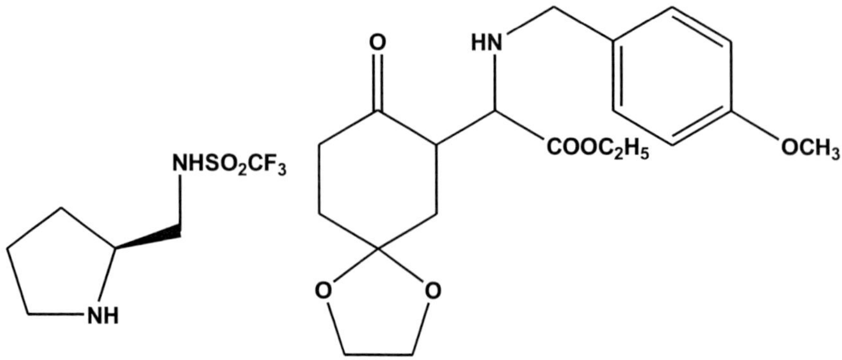

(66) **(67)**

5. NITROSATION

The nitrosation proceeds very rapidly with immediate evolution of carbon dioxide upon the addition of the nitrite.

Although 1,2-cyclohexanedione monoxime **(68)** and its derivatives can be prepared by direct nitrosation of the corresponding monoketones. They are also available from the nitrosation of ethylcyclohexanone-2-carboxylate **(1)** [89-91].

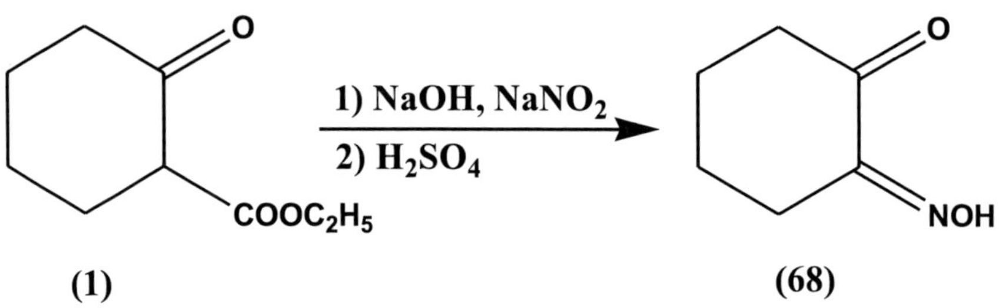

(1) **(68)**

Cyclic β-keto esters are usually cleaved to α-oximino diesters by nitrosation in the presence of alkali alkoxides. 2-Carboethoxy-4-methylcyclohexanone **(69)** is converted into diethyl α-oximino-methyladipate **(70)** in 25-30% yield by the action of nitrous fumes and sodium ethoxide, but almost twice this yield results from the use of ethyl nitrite and sodium ethoxide [92].

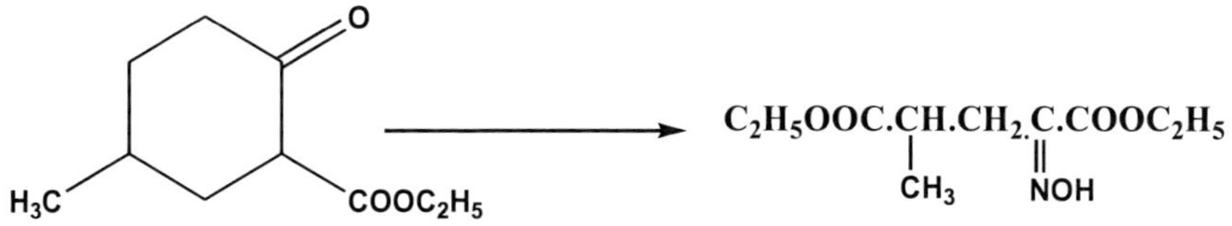

(69) (70)

6. THE JAPP-KLINGEMANN REACTION

When the Japp-Klingemann reaction is applied to a cyclic β-keto esters the ring is opened in the second stage of the process; as e.g the reaction of ethyl cyclohexanone-2-carboxylate **(1)** with benzenediazonium salt to form **(71)** [93, 94].

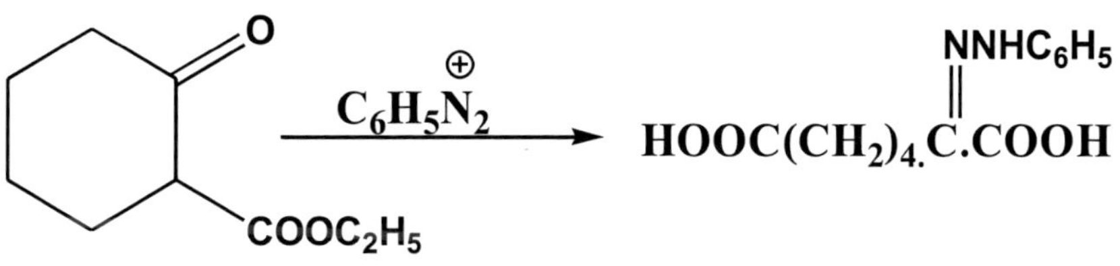

(1) **(71)**

Diethyl 1,4-cyclohexanedione-2,5-dicarboxylate **(7)**, couples with the diazonium compounds to form 1-aryl-4-arylazo-3-carboethoxy-5-pyrazolones **(72)**. This was proved by Fcofilaktov et al. [95], and Yasuda [96].

In this particular reaction there is no established evidence yet to show in which step the cleavage of the cyclohexane ring takes place. The probable scheme is illustrated as follows by Yasuda [96]:

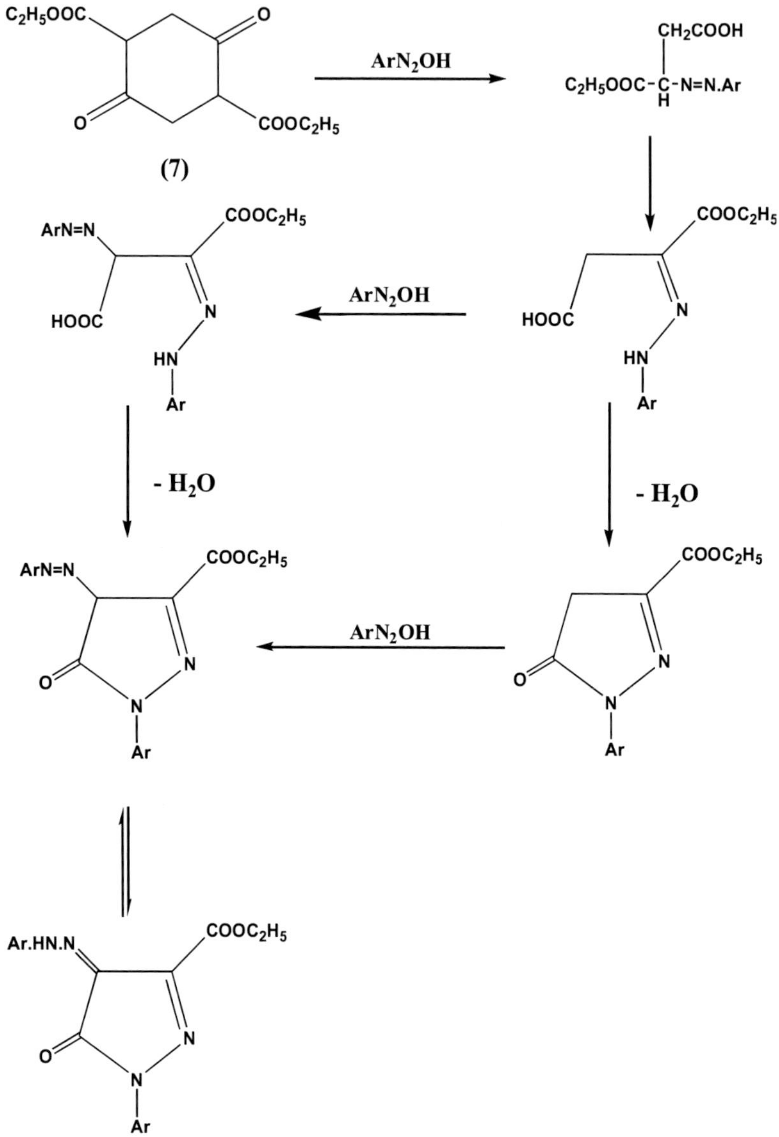

(7)

(72)

In contrast to the above reaction , Metwally et.al [97] reported that, when **(7)** was treated with dilute sodium hydroxide for overnight, before coupling with diazonium salt, 1,2,4,5-cydohexaterone-2,5-bis (arylhydrazones) **(73)** were obtained.

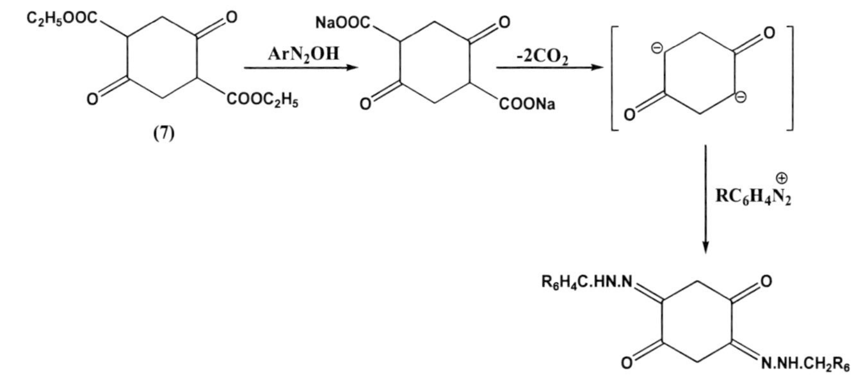

(7)

(73)

Metwally et al. [98] reported the formation of **(74)** and **(75)** when **(37)** was subjected to Japp-Kingemann reaction in alkaline medium.

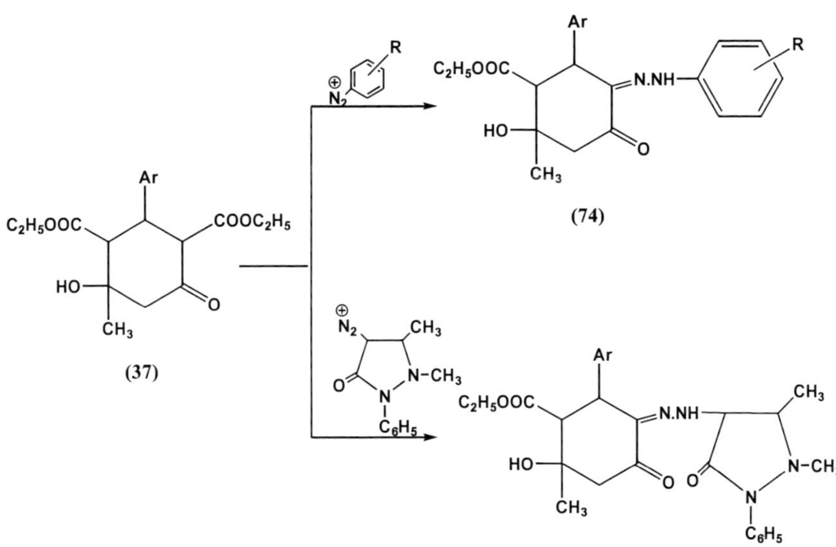

(37)

(74)

(75)

7. Reaction with Amines

The reaction between diethyl 1,4-cyclohexanedione 2,5-dicarboxylate **(7)** and amines to give diethyl 2,5-diamino-1,4-dihydroterephthalates **(76)** was first reported by Baeyer in 1886 [99].

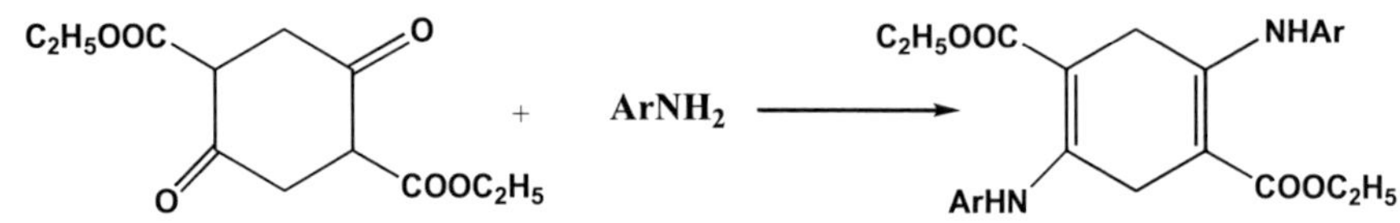

(7) (76)

The condensation of **(7)** with aromatic amines is affected by boiling in alcohol containing glacial acetic acid or hydrochloric acid under nitrogen atmosphere [100, 101]. This condensation can also be carried out in Dowtherm A (an eutectic mixture of 23.5% biphenyl and 76.5 biphenyl oxide) using p-toluenesulphonic acid [102], or an arylamine hydrochloride as catalyst [103].

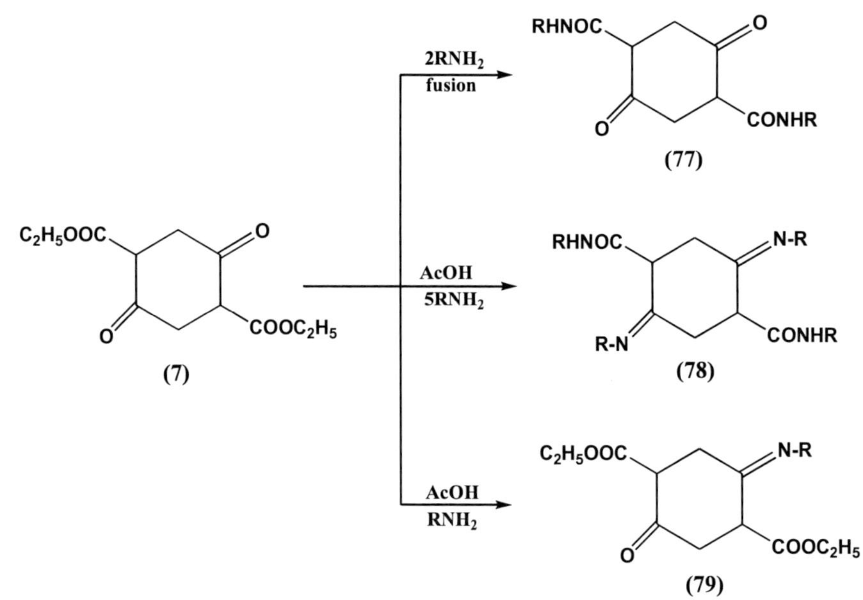

Metwally et al. reported that diethyl 1,4-cyclohexanc-dione-2,5-dicarboxylate **(7)** undergoes condensation with amines to give the 2,5-dicarboxamids **(77)**, 2,5-bis (phenylimino)-1,4-cyclohexanedicarboxanilides **(78)** 61 and 4-phenylimino-cyclohexanone-2,5-dicarboxylate **(79)** [104].

Benzoxazepiinone of the type **(80)** was obtained from **(7)** and 0-aminophenol in two steps as follows [105]:

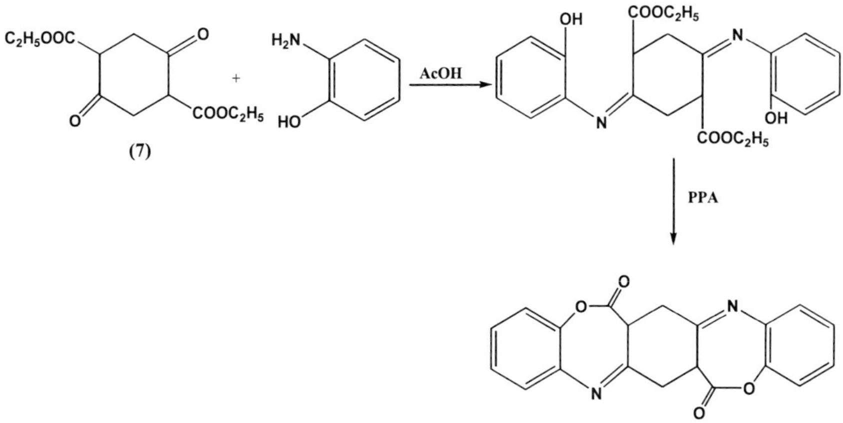

(80)

In 1968 Kimura [106] reported that **(7)** was condensed with m-phenylenediamine and the resulting polyamine was oxidized and cyclized to form a polyquinolone **(81)**. Cyclization of the polyamine without oxidation produced dihydroquinoloquinolone **(82)**.

Similar condensation of **(7)** with 4,4-diaminodiplhenyl ether followed by oxidation and cyclization yielded a poly (oxyquinacridone) **(83)**.

Later Higashi et al. [107]. reacted **(7)** with aliphatic diamines to obtain polymers which formed tough films of higher molecular weight than those reported by Kimura. Higashi, et al. [107], have shown the reaction between **(7)** and amines occurs in successive steps.

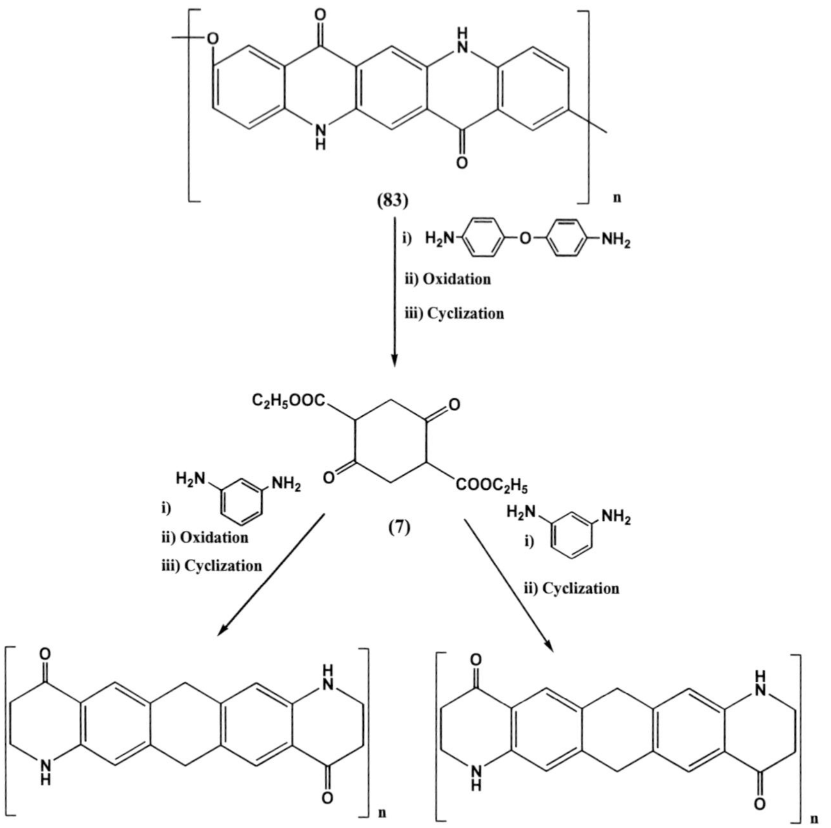

(83), (7), (81), (82)

At room temperature, an unstable ammonium salt **(83)** is formed which on heating to 50-70°C afforded mono amino-dihydroterephthalate ammonium salt **(84)**.

At temperature over 100°C, **(83)** and **(84)** are converted to the diaminodihydroterephtthalate **(85)**. Oxidation of **(85)** or **(86)** to the aromatic moiety, **(87)** takes place if R in **(85)** and/ or **(86)** is aryl.

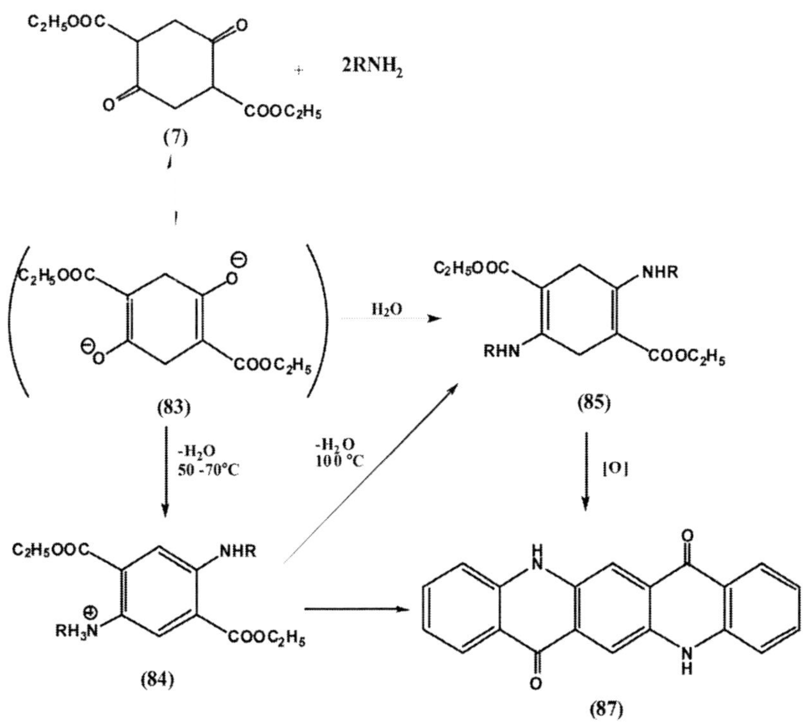

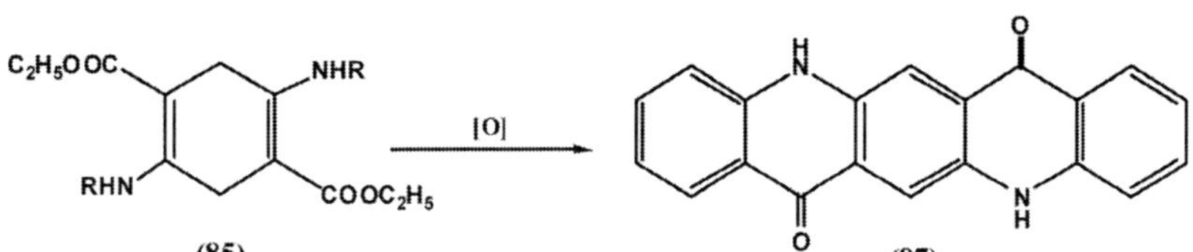

Moore, et al. [108] reported that **(7)** reacted with benzylamine to give diethyl 2,5-bis (benzylamino)-3,6-dihydroterpthalate **(88)** which was oxidized to the fully aromatic derivative diethyl 2,5-bis (benzylamino) terephthalate **(89)**.

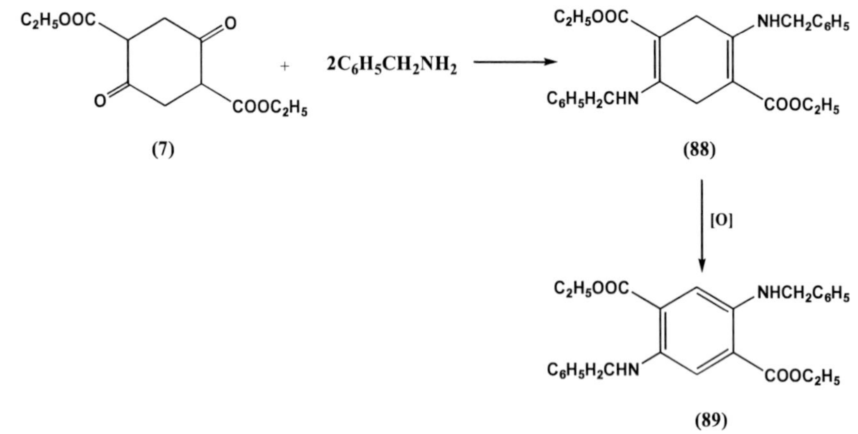

Polycondensation of **(7)** with benzylic diamines gave poly (amino-dihydroterephthalate (**90**; Z = arylene), which was oxidized to the fully aromatic form **(91)** [108].

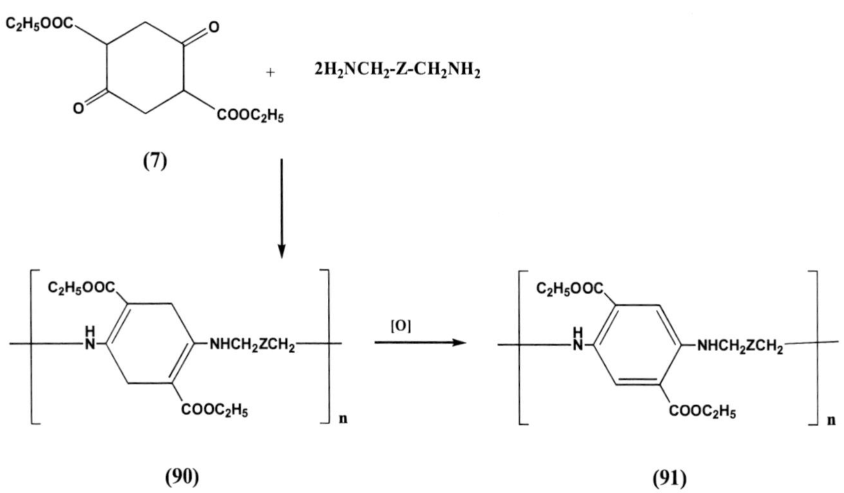

Yokoyama, et al. [109], reported that the reaction of **(7)** and 2-aminopyridine in the presence of a proton acid yields a mixture of 6,7,14,15-tetrahydropyrido[2,1-b] pyrido [1,2:1,2] pyrimido [4,5-g] quinazoline-7,14-dione **(92)** and 7-ethoxycarbonyl-6,9-dihydro-8-hydroxy-10H-prido[2,1-b] quinozolin-10-one **(93)**. Similar results are also obtained from the reaction of 2,5-diamino-1,4-bis (ethoxycarbonyl)-1,4-cyclohexadiene (**94**; R=H) and 2-amino pyridine [110].

The formation of these products is consistent with initial slow condensation of 2-aminopyridine with **(7)**, or its anion-exchange reaction with (**94**) followed by rapid intramolecular cyclization to give **(92)** and **(93)** [112]. Dehydrogenation of **(92)** gives 7,15-dihydropyrido [2,1-b] pyrido-[1,2:1,2]-pyrimido [4,5-g] quinazoline-7,14-dione **(95)** [109].

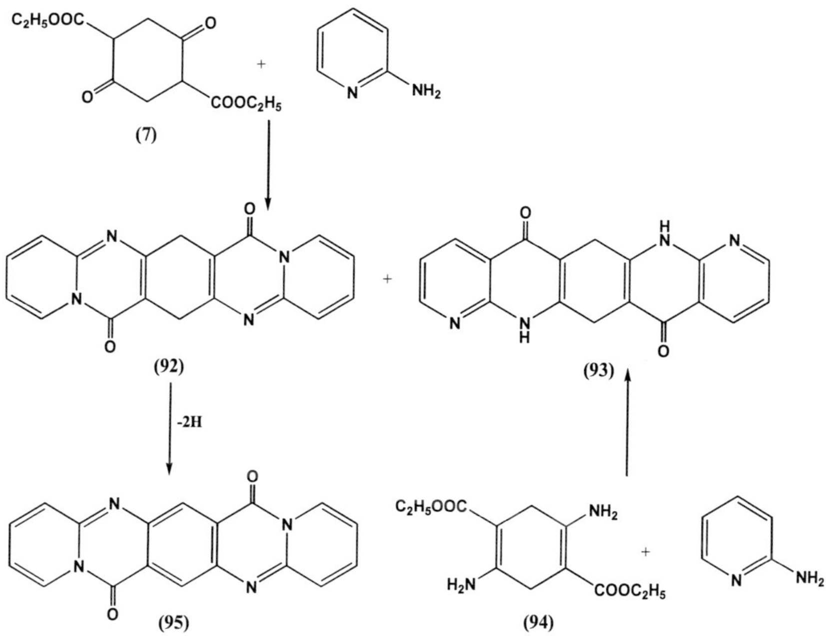

Quinacridone of the type **(99)** was obtained by Ehrich and Jaffe [111] by several methods. The first method includes the condensation of **(7)** with p-aminoethylbenzoate to give dialkyl 2,5-bis (4-carboethoxyanilino)-3,6-dihydroterephthalate **(96)**, which was cyclized to the 6,13-dihydroquinacridone **(97)** in boiling Dowtherm (an eulectic mixture of 23.5% biphenyl and 76.5% biphenyloxide) and then oxidized and hydrolyzed, or by condensing **(7)** with p-aminobenzoic acid to give the 4-carboethoxy analog of (**96**), which was either oxidized and hydrolyzed and then cyclized with polyphosphoric acid, or was first cyclized in boiling Dowtherm and then oxidized. The second method includes the hydrolysis of 2,9-bis (trifluoromethyl) quinaacridone **(98)** in sulphuric acid, to give **(99)**.

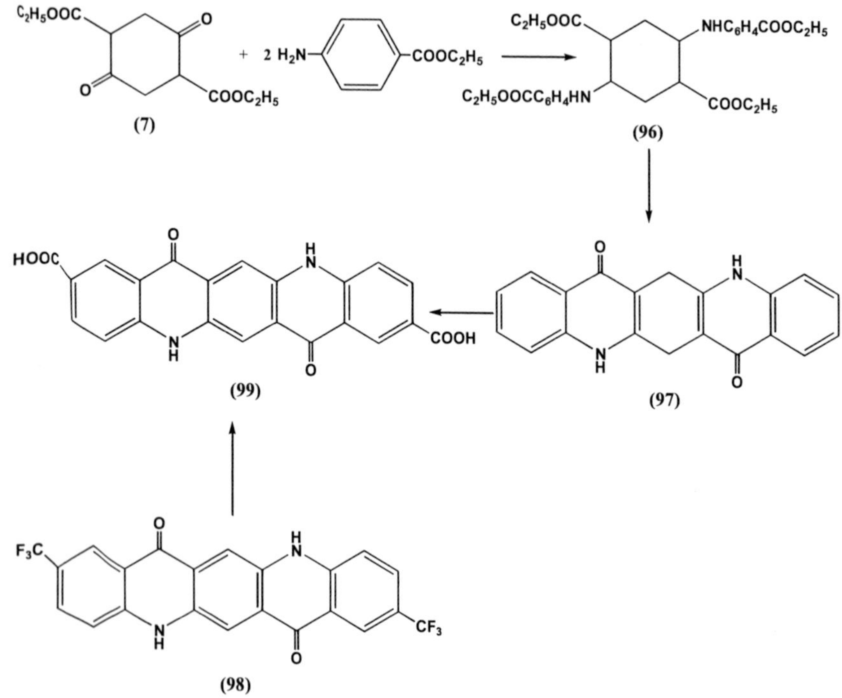

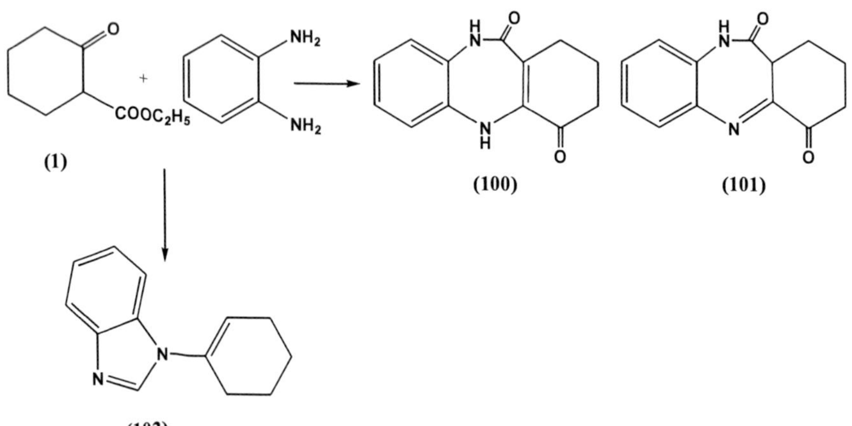

Ethyl cyclohexanone-2-carboxylate **(1)** was treated with o-phenylenediamine to give the diazepins **(100)** and **(101)** respectively, while Rossi et al. [112] found that N-cyclohexenyl benzimidazole **(102)** was prepared from the condensation of **(1)** with o-phenylene-diamine.

On the other hand, Ege-Guenter [113] reported that the condensation of **(1)** with o-,m-,p-diaminobenzene and 1,3,5-triaminobenzene produced imino bis (cyclohexenecarboxylates) which underwent cyclization in Ph_2O to give **(103)**, **(104)**, **(105)** and **(106)** respectively.

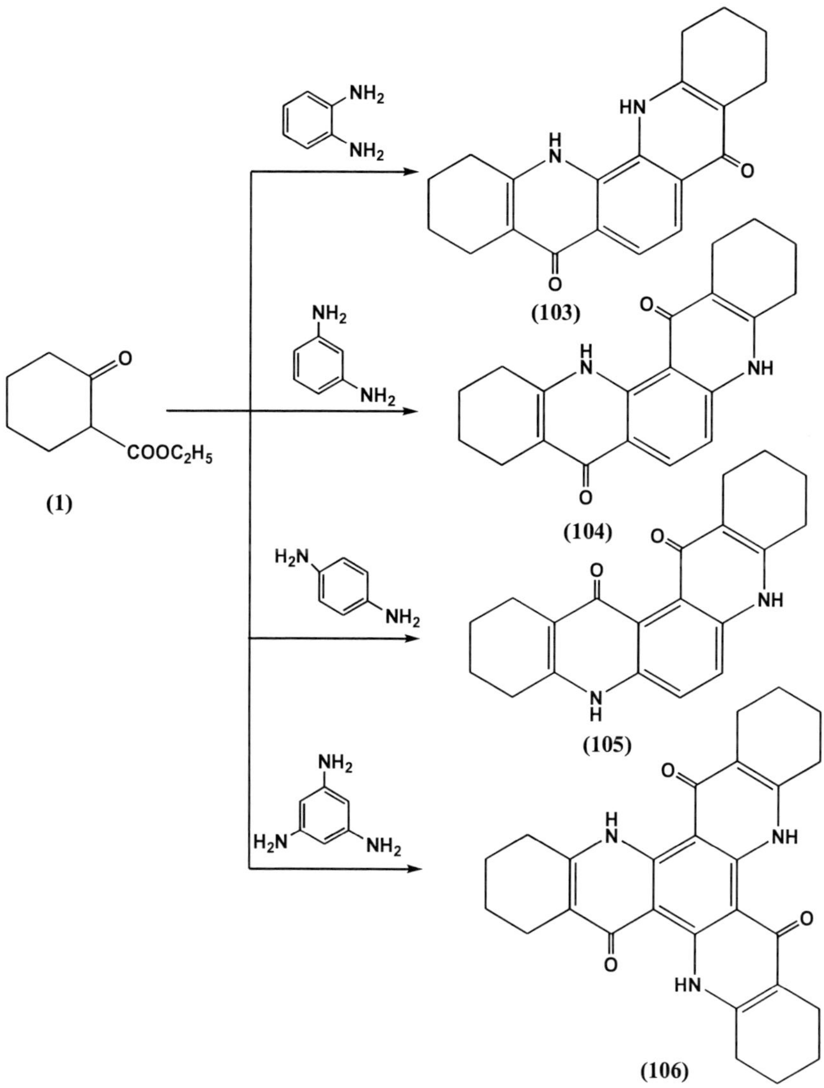

(1) (103) (104) (105) (106)

The reaction between **(7)** and o-phenylenediamine in hot acetic acid has been reported to give 6, 6a, 8, 14, 14a, 16-hexahydrobenzo-[1,2-b : 4,5-b] bis

[1,5] benzodiazepine-7,15-dione **(108)**. The intermediate dianilide **(107)** has been obtained upon fusion of **(7)** with o-phenylenediamine at 140°C, which underwent cyclization to give **(108)** upon treatment with hot acetic acid [97].

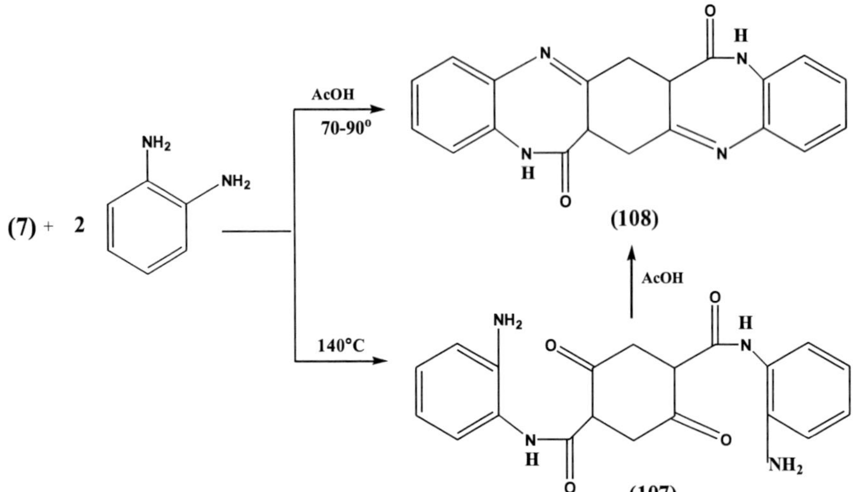

Catalytic enantioselective α-amination reactions of β-keto esters and β-keto lactones with dibenzyl azodicarboxylate are catalyzed by cinchona alkaloids, affording the products **(109)** [114].

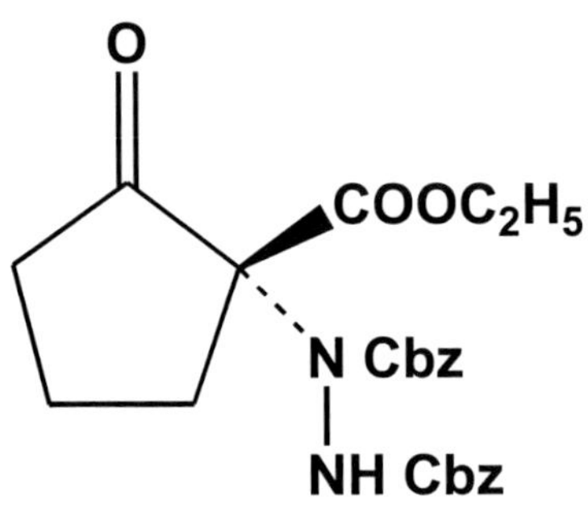

(109)

2,2,6,6-Tetramethyl-4-piperidinyl acetoacetate was condensed with $MeOCH_2CH_2NH_2$ in refluxing toluene containing HOAc to give 89% $MeOCH_2CH_2NHCMe:CHCO_2Q$(Q=2,2,6,6-tetramethyl-4-piperidinyl) [115a].

A highly efficient asymmetric amination of β-keto esters was achieved under phase transfer conditions using chiral quaternary ammonium bromide as a catalyst. The amination products were obtained in quantitative yields with up to 97% ee. One of the amination products represents a key intermediate for the preparation of aldose reductase inhibitor AS-3201[115b].

8. Reaction with Hydrazines and Acid Hydrazides

2-Carboethoxy cyclohexanone **(1)** was reported to give cyclohexenylcarboxylic and **(110)** [116], through their reaction with hydrazine hydrate.

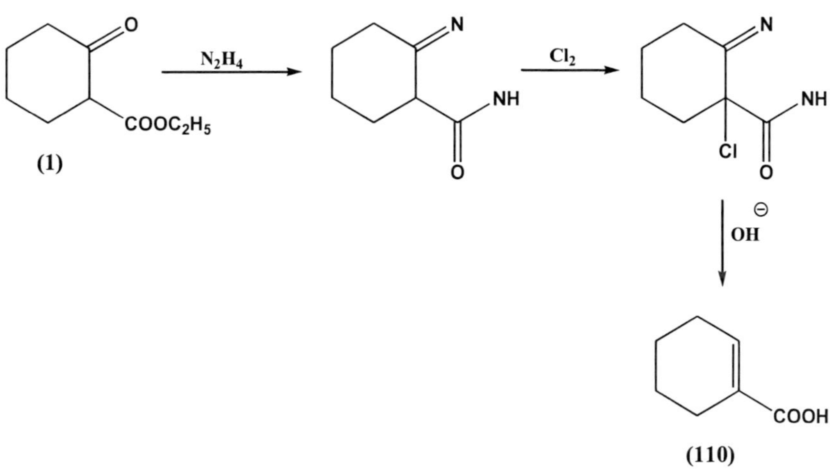

(1)

(110)

Lal and Bhaduri [11] reported that 3-aryl- 2,4- dicarboethoxy (or dicarbomethoxy) –5-methylcyclohexanones **(37)** reacted with hydrazine or substituted hydrazines to give 4,5-trans-2 (h)-alky1(or ary1)-4-ary-5-carboethoxy (or carbomethoxy)-3,6- dihydroxy-6- methy1-4,6,7-tetrahydroindazoles **(111a)**.

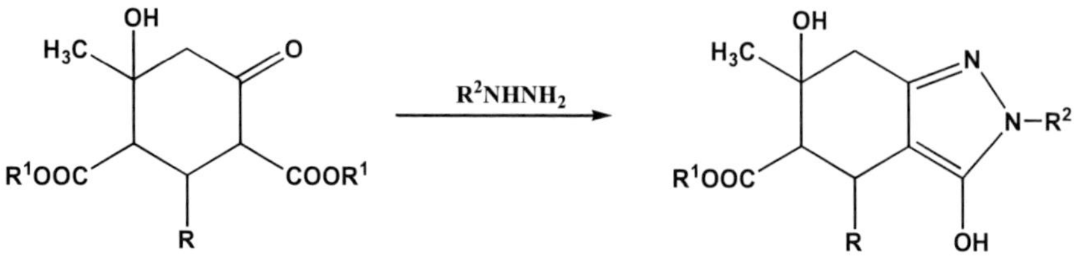

(37) (111a)

Metwally et al. [117] reported that condensation of **(28)** with hydrazines in acetic acid to give indazoles **(111b)**.

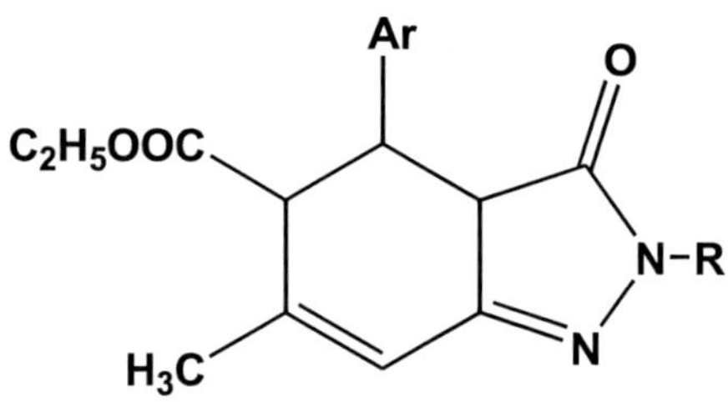

(111b)

Metwally et a1. [97]reacted **(7)** with two moles of hydrazine hydrate in ethanol, 1,2,3a,4,6,7a,8-hexahydro-2,6-bis(substituted) benzo [1,2-c:4,5-c] dipyrazole-3,7-diones **(112)** was obtained in good yield.

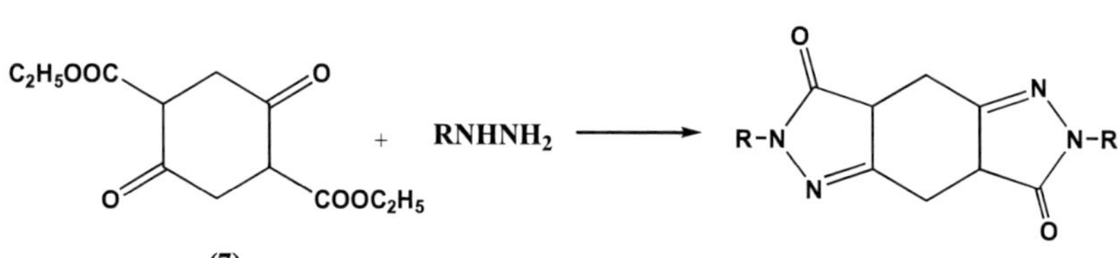

(7) (112)

When the same reaction was carried out using acetic acid, diethy1-2,3-diazabicyclo[2,2,2] octa-(7), 4-dione-5,7-dicarboxylate **(113)** and diethy1 2,5-dioxo-1,4 –cyclohexanedicarboxylate–2,2-azine **(114)** were obtained [104].

Diethyl 1,4-cyclohexanedione-2,5-dicarboxylate **(7)** reacted with one mole of benzenesulphonehydrazide to give ethyl 3,3a,4,5,6,7-hexahydro -4,5- dioxo -2- phenylsulphonyl -2H- indazole -6- carboxylate **(115)** [104].

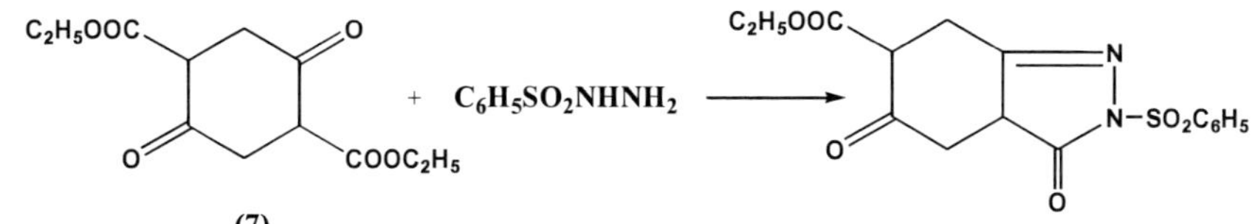

(7) (115)

Condensation of 4-phenylimino-cyclohexanone-2,5-dicarboxylate **(79)** [104] with hydrazine hydrate in ethanolic solution gave tetraethyl 2,2-azobis (5-anilinoterphthalate) **(116)**.

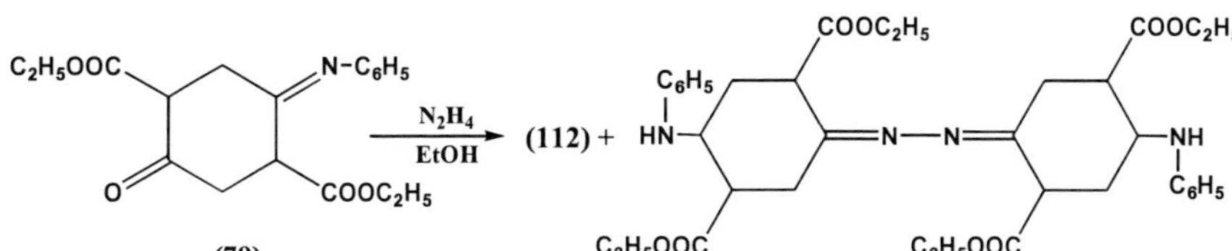

(79) (116)

Metwally et al. [118], reacted 2-pyridylhdrazine and / or 2-hydrazino-5-sulphobenzoic acid with **(7)** and isolated bis (aryl) benzo [1,2-c:4-c,]dipyazole-3,7-diones **(117a,b)**.

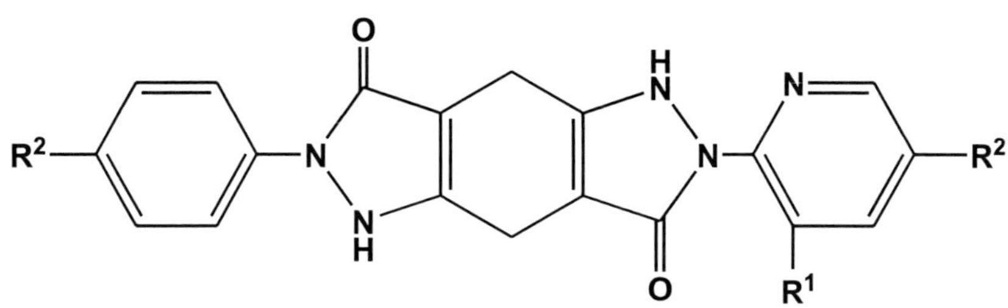

(117)

a, X=N, R^1=R^2=H
b, X=CH, R^1=COOH, R^2=SO_3H

The condensation between **(7)** and acid hydrazides was reported to give the hydrazones **(118)** [79].

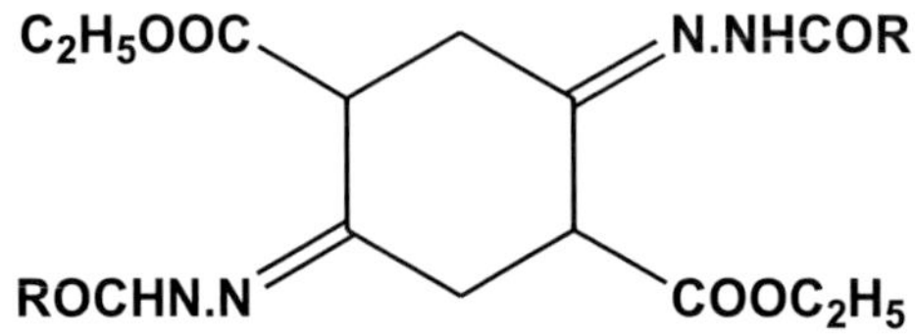

(118)

Metwally et al. [118], reported that condensation of **(7)** with N,N-diphenylsulphonylhydrazide in ethanol gave diethyl 2,5-dioxo-dioxo-1,4-cyclohexanedicarboxylate-2,5-bis-(N,N-diphenylsulphonyhydrazone) **(119)** in a good yield. Further treatment of **(119)** with 4% acloholic potassium hydroxide and acidification with concentrated hydrochloric acid afforded (**112**).

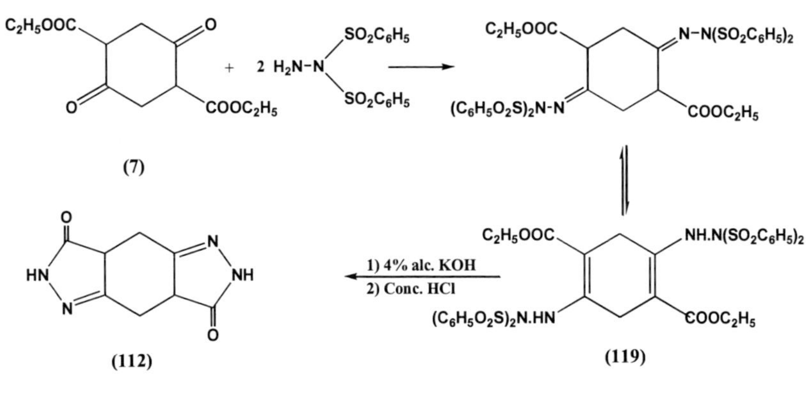

9. Reaction with Hydroxylamine

The condensation of diethyl 1,4-cyclohexanedione – 2,5- dicarboxylate **(7)** and hydroxylamine hydrochloride in pyridine has been reported to give 3a, 4, 7a, 8- tetrahydro-3h, 7h- benzo[1,2-c:4,5-c'] diisoxazole-3,7-dione **(120)** [104].

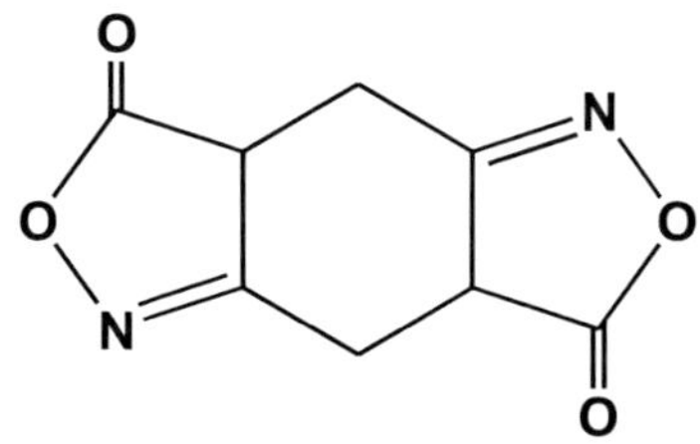

(120)

Metwally et al. [117] reported that 3-aryl-2,4-dicarbethoxy-5-methyl cyclohexanones (**37**) was condensed with hydroxylamine hydrochloride in presence of sodium acetate to give ethyl1,3,3a,4,5-tetarahydro –6-methyl-3-oxo-4- (aryl)-2,1-benzisoxazole-5-carboxylate **(122)**. When the same reaction was carried out using ethanol instead of acetic acid the oximes **(121)** was obtained, which underwent cyclization with acetic acid and sodium acetate to give **(122)**.

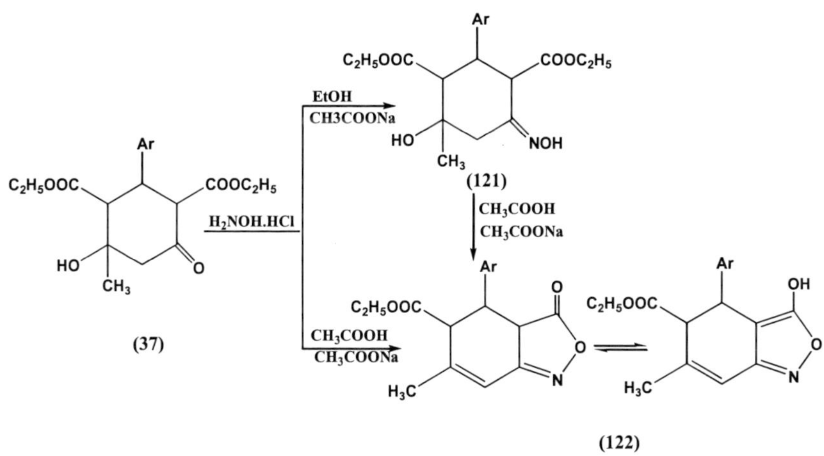

It has been reported that, the reaction between **(1)** and hydroxylamine under basic condition followed by acidification normally afforded a mixture of isoxazolin –5-ones **(123)** and 3- hydroxy isoxazole **(124)** [119].

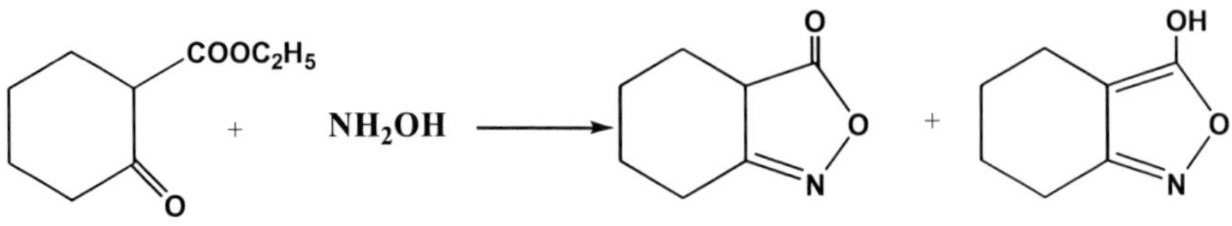

(1) (123) (124)

10. Reaction with Urea and Thiourea

The reaction of **(7)** with urea and /or thiourea in ethanolic sodium ethoxide solution gave 4a,5,9a,1tetrahydropyrimido [4,5-g] quinazoline-2,4,7,9 (3H, 8H)tetrone **(124)** [107] and 4a,5,9a,10- tetrrahydro-2,7-dithiopyrimido [4,5-g]quinazoline-2,4,7,9(3H,8H)-tetrone **(125)** [105] in good yield .

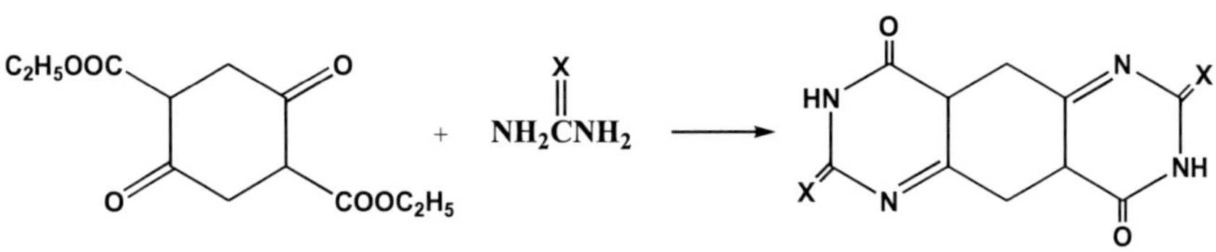

(7)

(124, X= O)
(125, X= S)

Taschner [120] reacted **(7)** with thiourea and obtained diethyl 2,5-dihydroterephthalate **(126)**. Thiourea in this case acts as an oxidizing agent, this oxidation is the first reported case of an oxidative action of thiourea.

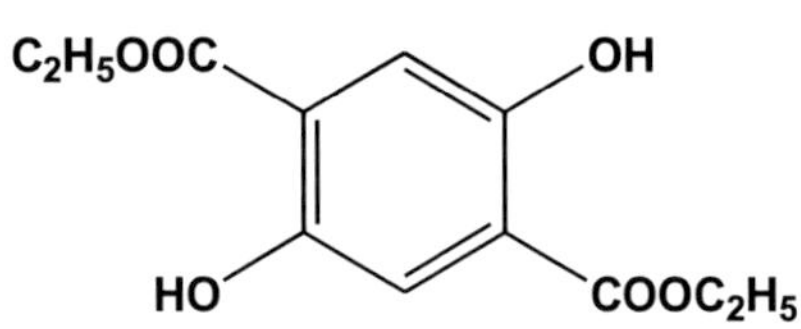

(126)

Metwally et al. [117] reported that, the condensation of **(37)** with thiouea in acidic medium gave the thiazines **(127)**, while the condensation in ethanolic sodium ethoxide afforded the thiouracils **(128).** Reaction of **(128)** with monochloroacetic acid in water for about 2 hours resulted in the formation of quinazolines **(129)**.

On the other hand, condensation of **(37)** with urea, led to the formation of uracils **(129)** [117].

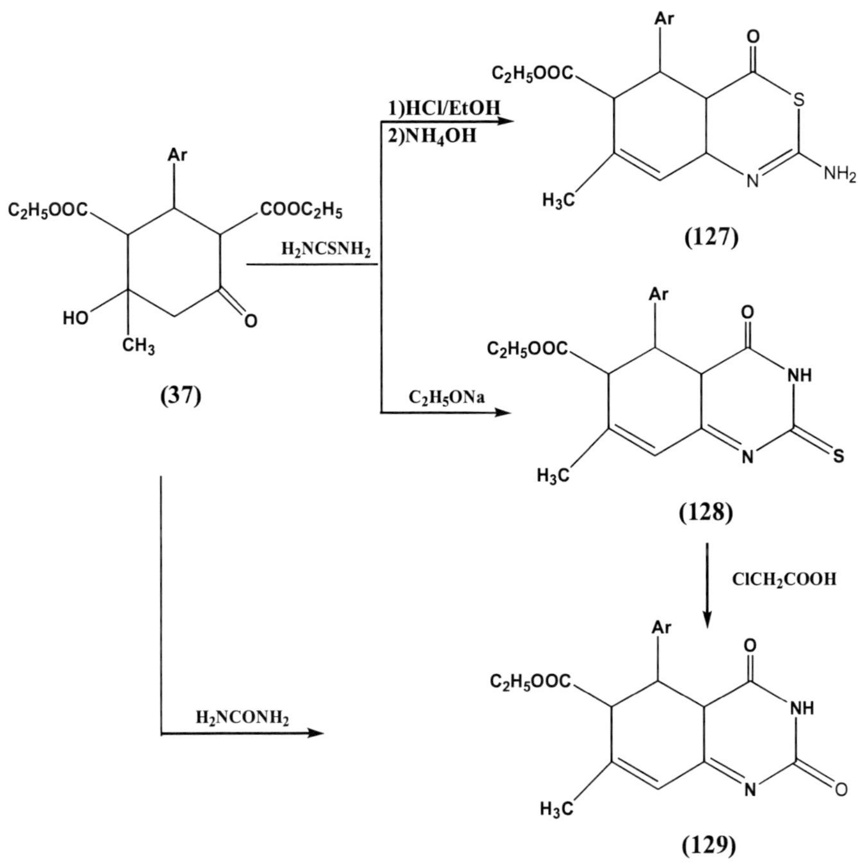

Reaction of **(1)** with thiourea gave 1-cyclohexyl-2-thiouracil **(130)** [120]. o-Methylthiourea reacts smoothly with **(1)** to give the pyrimidinone **(131)** [121].

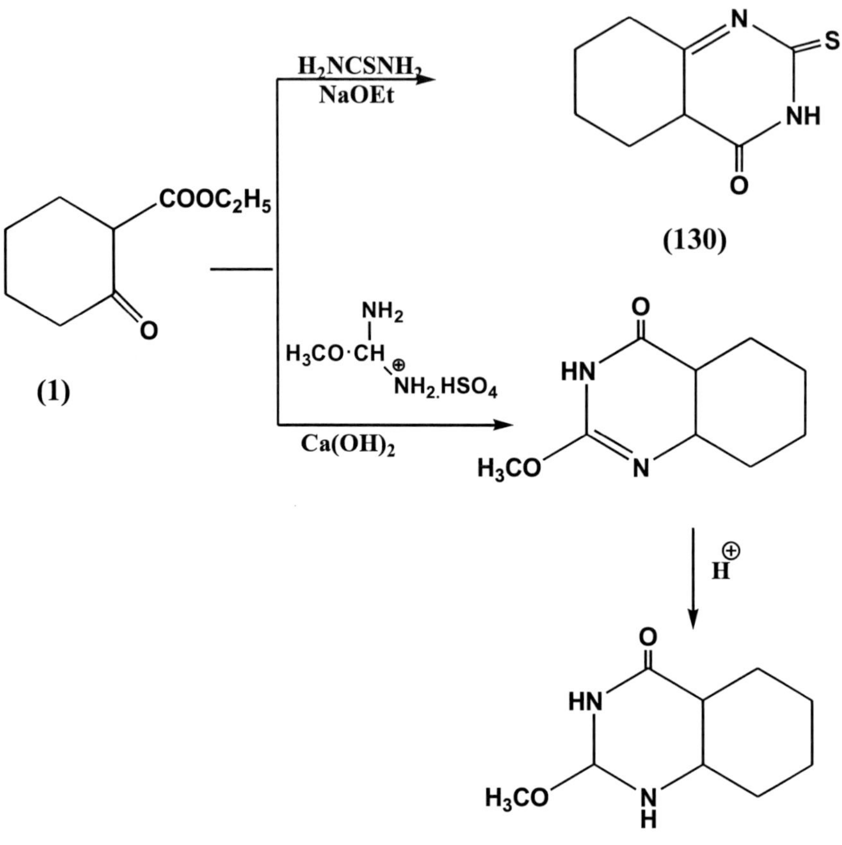

(1) (130) (131)

11. Reaction With Guanidines

It has been reported that 2-carboethoxycyclohexanone **(1)** reacted with guanidine carbonate to give aminopyrimidinone **(132)** [122].

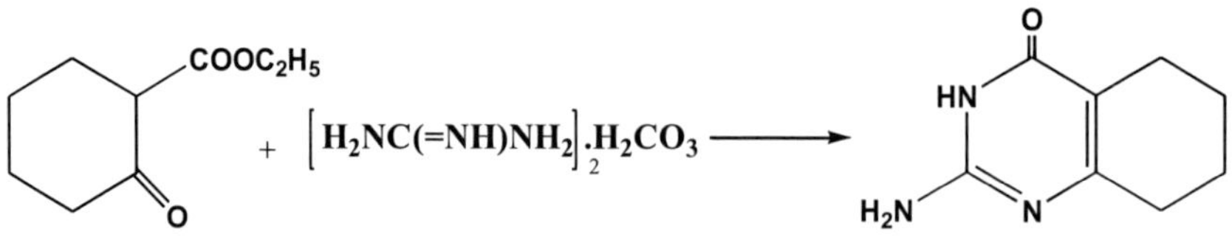

(1) (132)

On the other hand, a new pharmacologically useful drug ;2,3- diamino-4(3h) pyrimidinone **(134)** was prepared by condensing **(1)** with amino guanidine to give **(133)** which underwent allylic bromintion with NBS to give **(134)** [123, 124].

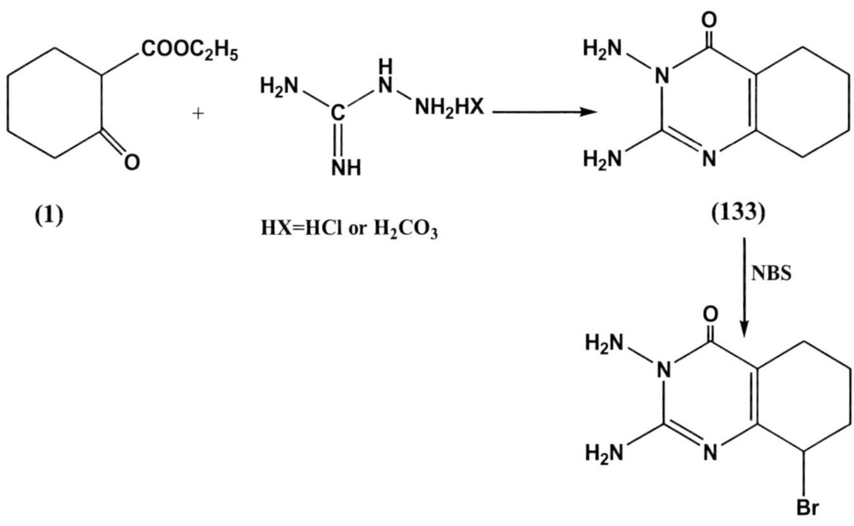

(1) **HX=HCl or H_2CO_3** **(133)**

(134)

Metwally et al. [118] reported that **(7)** reacted with guanidine hydrochloride to give 3,8-diamino-1,6- dihydroxypyrimidine [4,5-g]quinazolion dihydrocholoride **(135)** as a yellow powder.

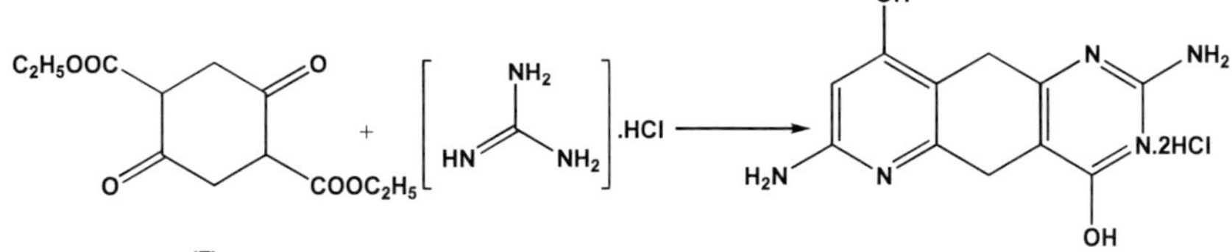

(7) (135)

12. Reactions With Aminoazoles

Stanovink et al. [125] has been reported that condensation of **(1)** with 3-amino-1,2,4-triazole in acidic medium produced **(136)**.

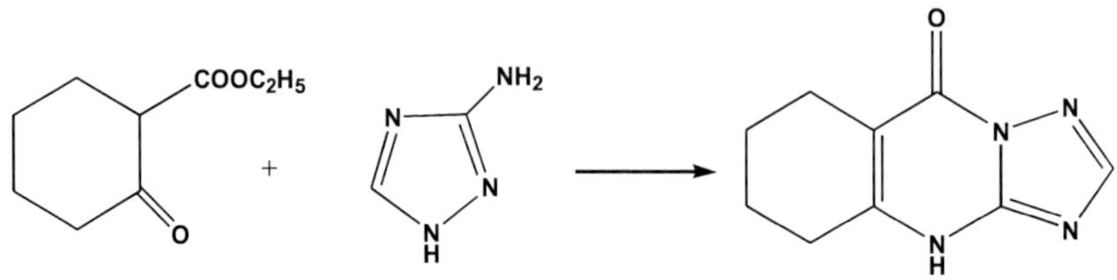

(1) (136)

While, Antaki and Petrow [126] studied the formation of 4:11-dihydro-4-keto-2:3-cyclohexeno-l-11-diaza-9-oxafluorene **(137)** via the condensation of **(1)** with 2- aminobenzoxazole.

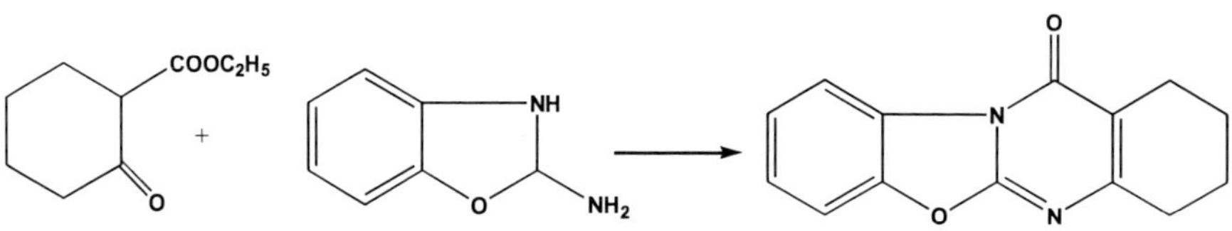

(1) (137)

The condensation of **(1)** with 2-aminobenzimidazole has been reported [126], to give 4,4a-dihydor–4-keto -2,3 cyclohexeno-1,4a,9- triazafluorene **(138)**.

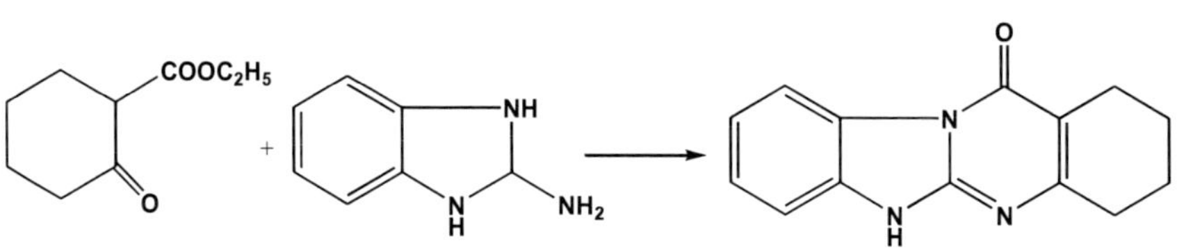

(1) (138)

On the other hand, Ward et al.[127] reported that pyrazolopyridine derivative **(141)** was reported via condensation of **(1)** with 4- aminopyrazole to give enamine **(139)** which underwent cyclization to pyrazolquioline derivative **(140)**.

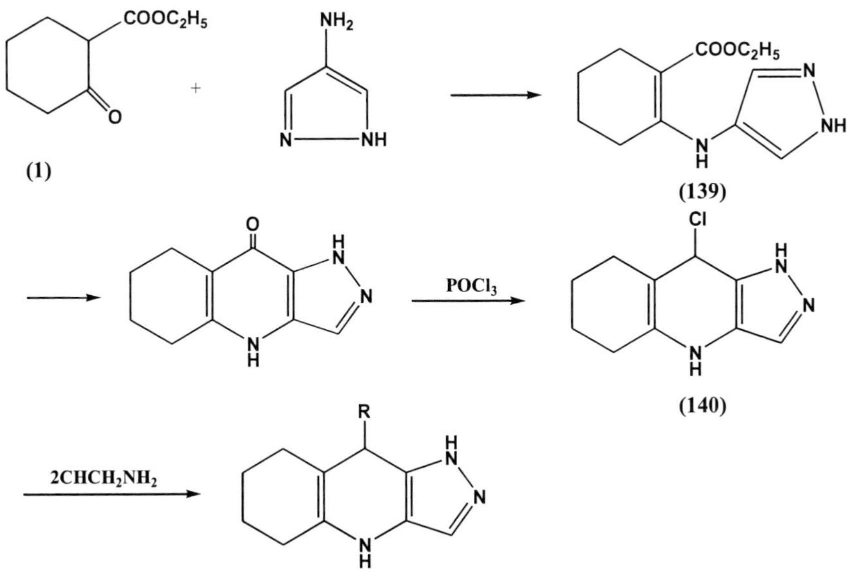

(1) (139) (140) (141)

R= $NH.CH_2CH{=}CHCH_2NH_2$

Condenstion of (**1**) with 3- aminoindazole in PPA at moderate temperature gave pyrimido (1,2-b)indazole [128] derivative **(142)**.

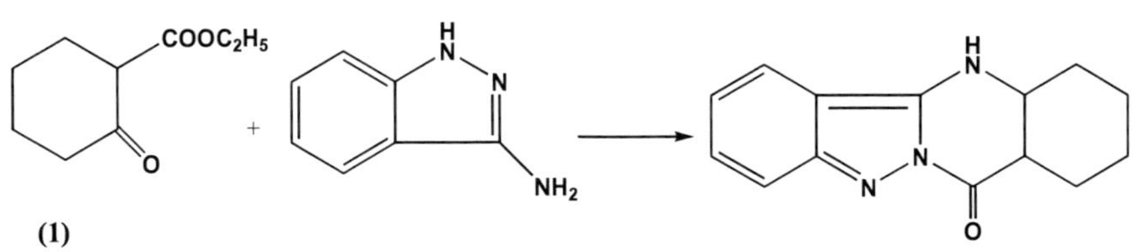

(1) (142)

Moreover, Kovluneko et al. [129] reported that pyrimidolone **(143)** and **(144)** had been prepared by cyclocondensation of **(1)** with 1- aminoisoindole.

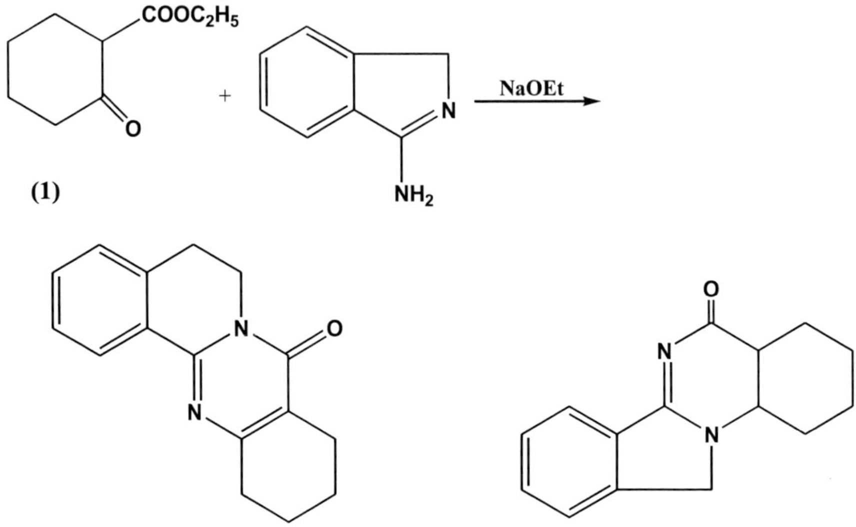

(1)

(143) (144)

Heinrich et al. [130] reported that pyrimidopyrimidine **(145)** was prepared by condensing (**1**) with 2-amino-3-ethoxycarbony1-1,4,5,6-tetrahydropridine.

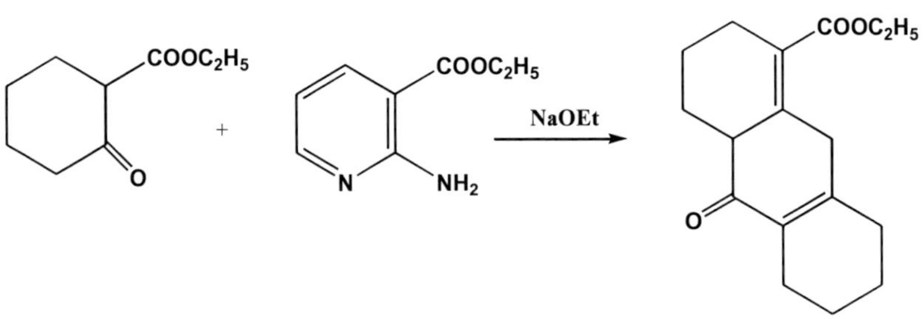

(1) (145)

Heinz et al. [131] utilized 2-amino-6-arylimidazo [2,1-b]-1,3,4-thiadiazole **(146)** to prepare imidazo [2,1-b]-1,3,4-thiadiazole-[3,2-a] pyrimidone **(147)** via the condensation with (**1**).

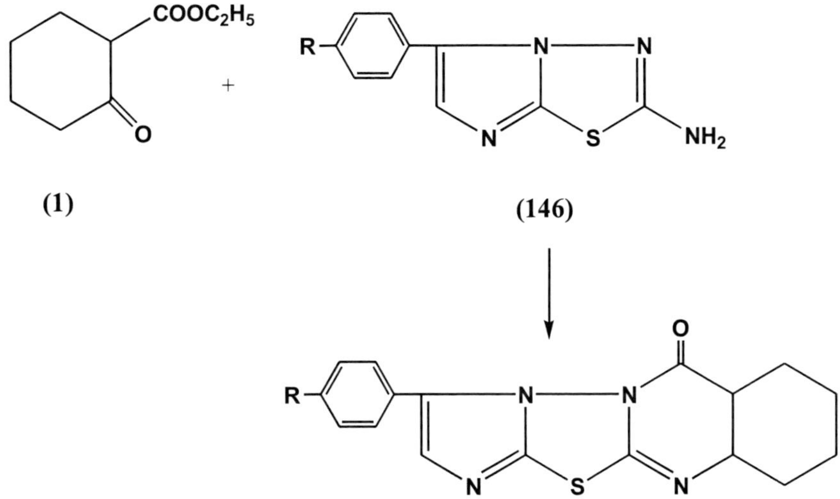

(1) (146) (147)

Metwally et al. [98] reported that condensation of 2- aminobenzimidazole,3-amino-1,2,4-thiazole, 5-aminotetrazole,1-pheny1-3-aminopyrazole with 3-ary1-2,4- dicarboethoxy-5-hydroxy-5-methy1 cyclohexanones **(37)** gave the linearly fused condensation products: pyrazolobenzopyrimidines **(148)**, pyrazolopyrimidines **(149)**, tetrazolopyrimidines **(150)**, pyrazolopyridines **(151)**, thiophenopyrimidine **(152)** and thiazolopyrimidines **(153)**.

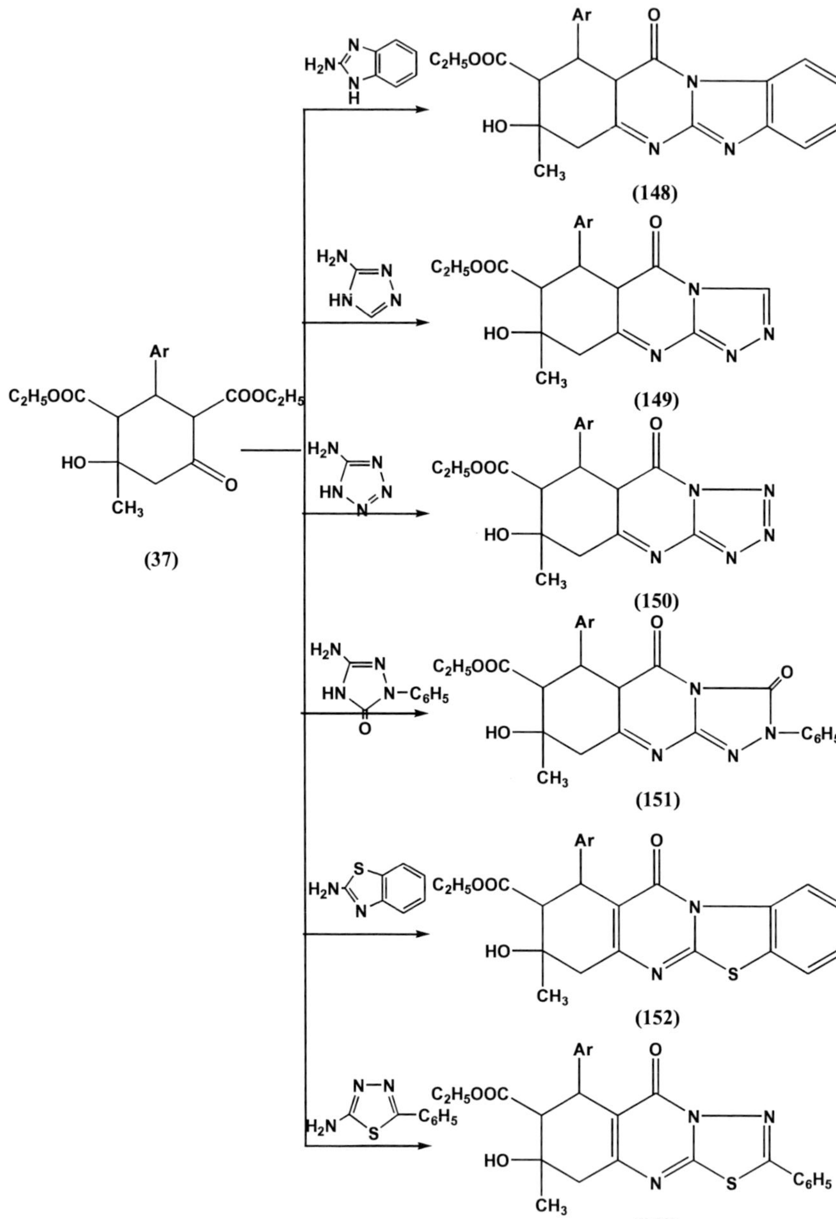

(37) (148) (149) (150) (151) (152) (153)

Metwally et al. [118] reacted **(7)** with 2- aminobenzoimidazole in ethanol-piperidine and obtained the anil drivative **(154)** which underwent cyclization with PPA to **(155)**. On the otherhand when the reaction between **(7)** and 2-aminobenzimidazole was carried out in PPA, **(155)** was obtained .

Similarly, the anil derivative **(156)** was obtained when **(7)** was condensed with 3-amino-1,2,4-triazole and 5-aminotetrazole in ethanol- piperidine. The product **(156)** underwent cyclization to **(157)** upon traetment with PPA. On the other hand when **(7)** was condensed with 3- amino-1,2,4-triazole and /or 5-aminotetrazole monohyderate in PPA, porduct **(157)** was obtained.

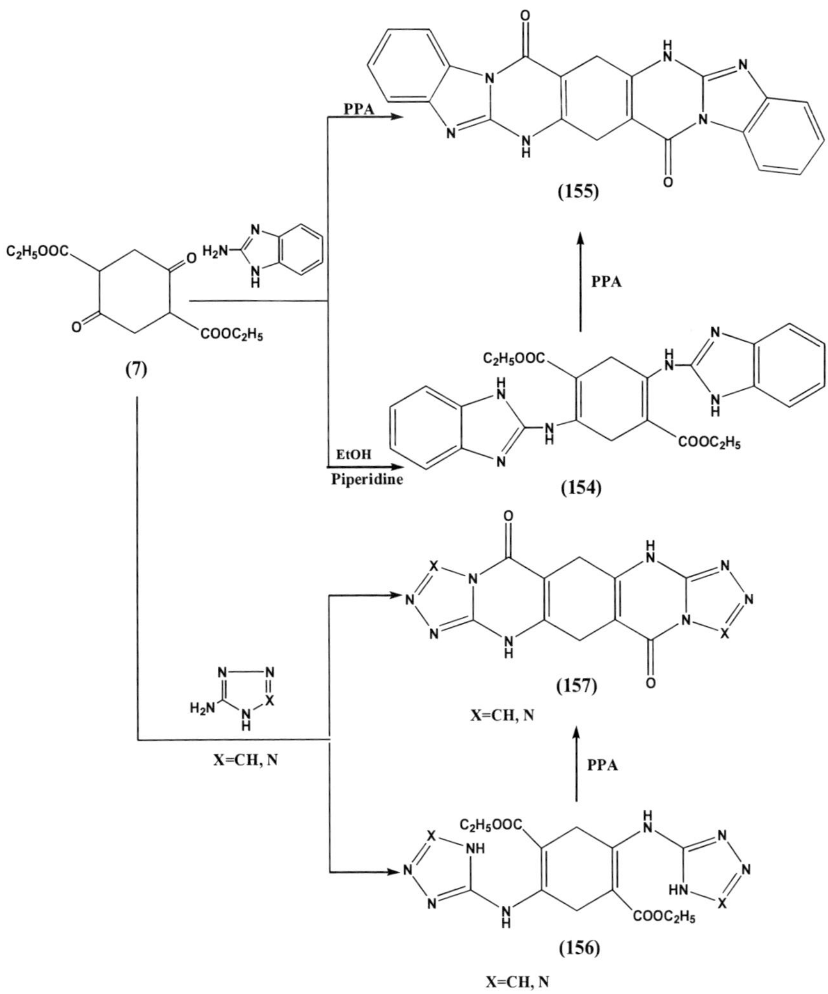

13. Reformatsky Reaction

Cyclohexanon-2- methyl-2-ethoxycarbonyl **(158)**, when treated with ethyl bromoacetate in presence of zinc metal gave the hydroxydiester **(159)** in a good yield.

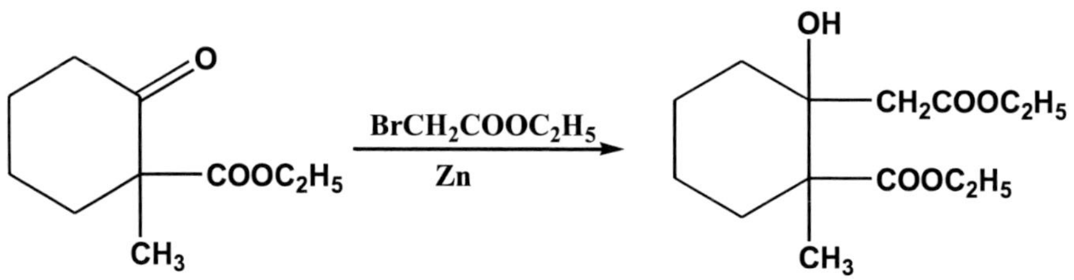

(158) **(159)**

14. Reaction With Glycols And Dithiols

Diethyl 1,4-cyclohexanedion-2,5-dicarboxylate **(7)**, when treated with ethylene glycol in the presence of a catalyst such as p-toluenesulphonic acid yielded 1,4-cyclohexanedione bis (ethyleneacetal) **(160)** [132].

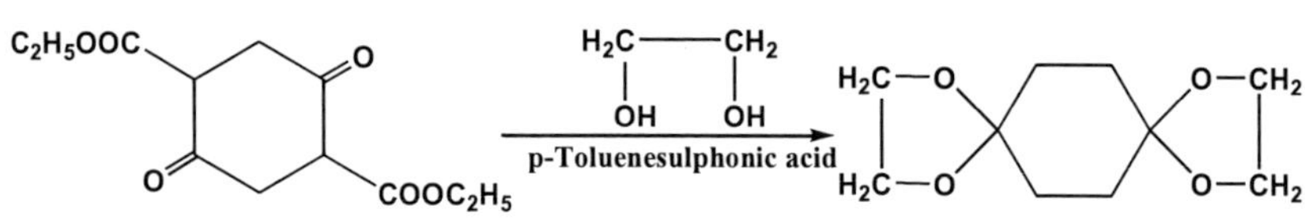

(7) **(160)**

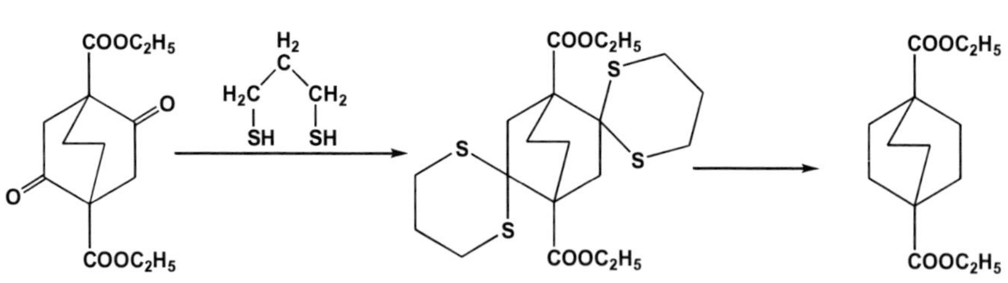

(161) **(162)** **(163)**

The reaction of β-keto ester, diethyl bicyclo [2.2.2] octanedione dicarboxylate **(161)** with dithiol under dehyderative conditions gives dithioketals **(162)** which are well known to cleave to the methylene derivative **(163)** upon treatment with Raney nickel [133].

15. Oxidation

Lampert et al. [134] reported that, ethyl-2-oxo-5,5-diethoxycyclohexane-carboxylate **(164)** undergoes a novel aromatization when heated with selnium dioxide giving **(165)**.

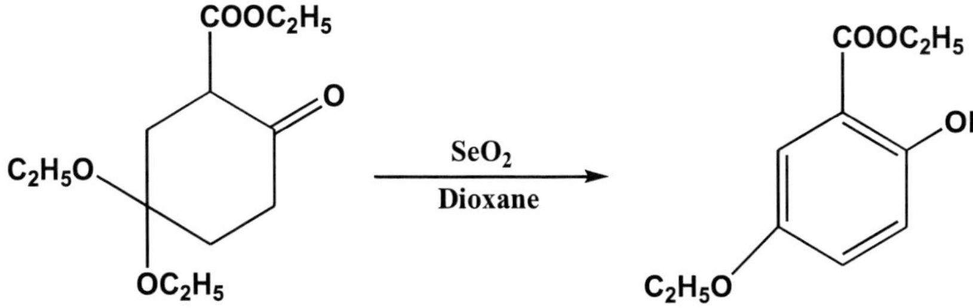

(164) **(165)**

The oxidation of cyclic β-keto esters **(166)** (n = 1, 2, 3; R = Me, Et) with O_2 was catalyzed with Mn (OAc).$4H_2O$ to give α-hydroxy derivatives **(167)**. On the other hand, Mn (OAc).$4H_2O$-catalyzed oxidation of $PhCOCH_2CO_2Et$ gave a low yield (6%) of Et cis-2,3-epoxy-3-hydroxy-3-phenylpropanoate [135].

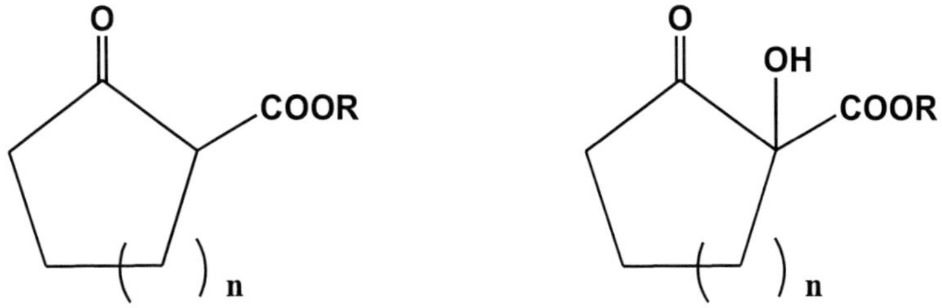

(166) **(167)**

An easy and convenient synthesis of 3-alkanoyl-5-(hydroxymethyl)-tetronic acids starting from malonate utilizing a direct route from cyclic β-keto esters to 1,3-dicarbonyl esters by phenylselenylation followed by oxidative cleavage is presented [136].

Oxidation of a non-aromatic or non-enonic ethylenic bond or of a non-conjugated cyclic ketones, in an inert solvent gave the corresponding epoxides, respectively lactone, using H_2O_2 as oxidant [137].

The radicals formed in Mn (III)-based oxidative free-radical cyclizations of alkenyl malonate esters and β-keto esters, e.g. **(168)**, can be trapped with sodium azide and Mn (III) to give cyclic and bicyclic azides, e.g. **(169)**, in 30-80% yield. Reduction of the azides gives bi- and tricyclic lactams, e.g. **(170)** [138].

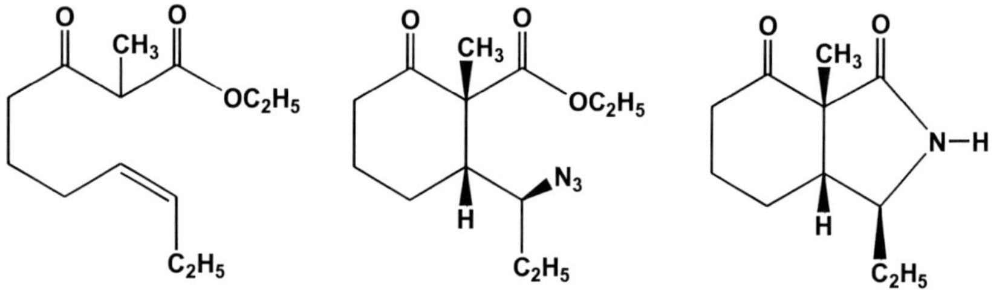

(168) **(169)** **(170)**

Selective oxidation of cyclic β-ketoester groups with KF-oxygen in Me_2SO provided α-hydroxy-β-keto-esters in fair yields. Application of the oxidation to 5-methoxycarbonyl-3-methoxycyclopent-2-enone furnished 60% (±)-kjellmanianone **(171)** [139].

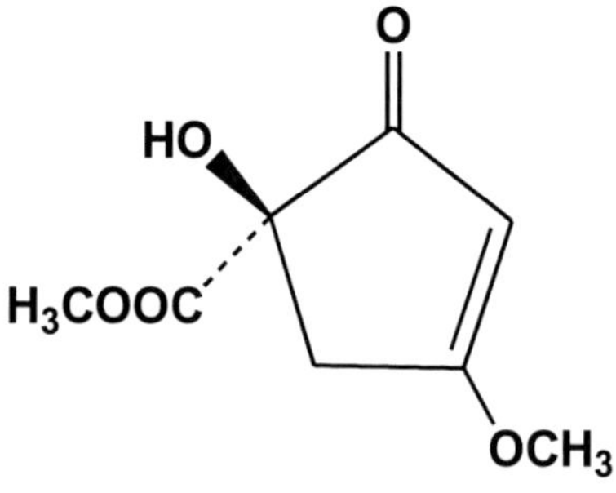

(171)

16. Reduction

In the cyclohexane analogue **(31)**, hydrogenation with platinum oxide in actic acid yields the cis-hydroxyesters **(172)**, while Raney nickel in methanol forms the trans –compound **(173)** [140].

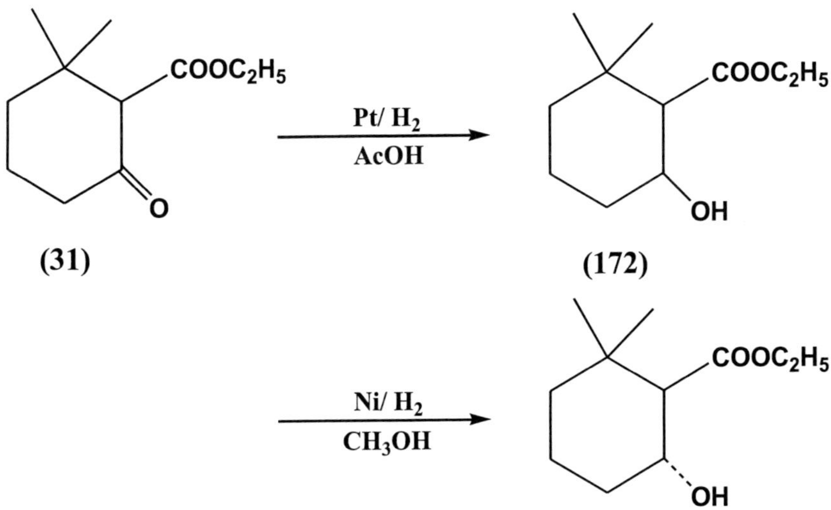

(31) **(172)** **(173)**

Hydrogenation of **(7)** over Raney Ni or Raney Co gave quantitative cyclohexandicarboxylates **(174)** (R= H; R =OH; R= OH; R=H) gave only 10 % conversions. Stopping the hydrogenation at 50% H_2 uptake gave the dihydro drivative **(175)** as well as both 69 isomers [141].

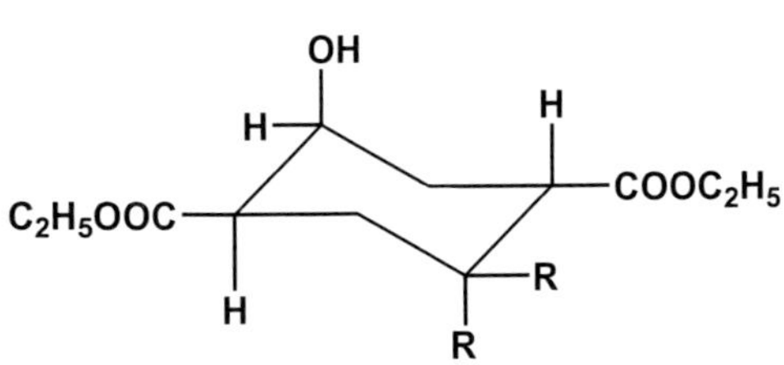

(174)

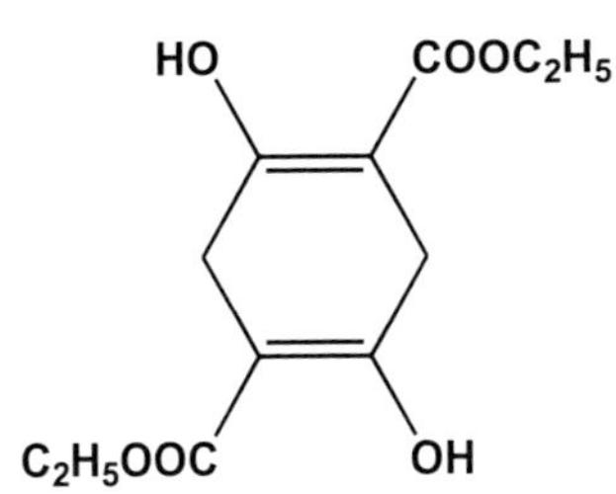

(175)

It was reportd that, the reduction of **(1)** with $LiAlH_4$ in ether gives rise to 1,3-diol **(176)** [142].

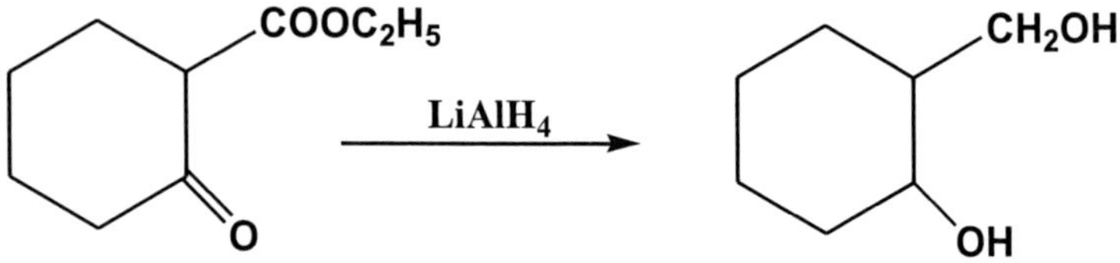

(1) **(176)**

On the other hand, the selective reduction [143] of **(1)** by sodium borohydride or by Raney nickel was reported to produce 2-hydroxycyclohexane carboxylate **(177)**, and the ratio of cis-trans isomer of the porduct obtained form the two reducing agents were determined and compared.

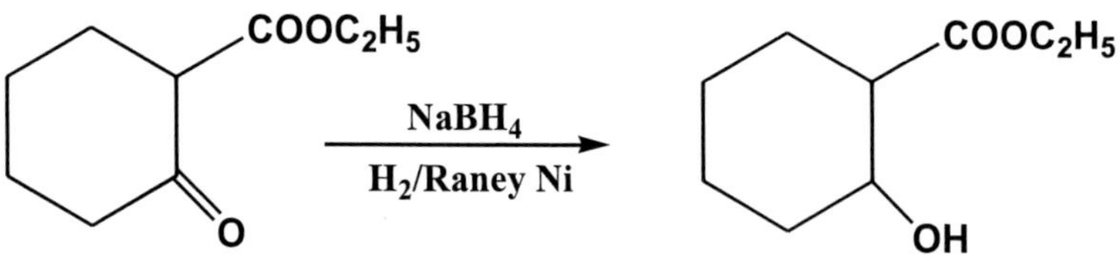

(1) **(177)**

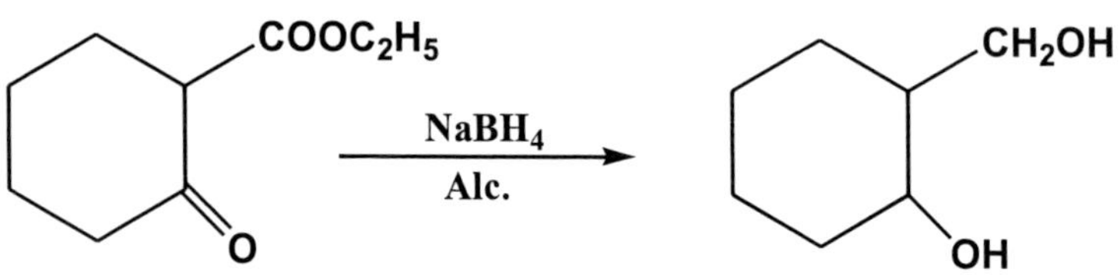

(1) **(176)**

On the other hand , it has been reported that **(1)** readily reduced by an optically active frementing mutant of saccharomyces cerevisiae and produce optically active (IR, 25)-(+)-2- hydroxycyclohexane carboxylate [144, 145]. Soai et al. [146] mentioned a chemoselective one step reduction of β- keto esters to 1,3-diols. Treating **(1)** with sodium borohydride in solvent system containg alcohol the 1,3- diol **(176)** was obtained.

Le guillanton [147] studied the effect of the nature of the electrolyte cation and the reaction temperature on the stereochemistry of the electrochemical reduction of **(1)** and found that the electrolyte cation (Et_4N^+, Li^+, Mg^{+2}) has no effect on the product ratio and gave 45, 81 and 90% cis **(177)**; 55,19 and 10% trans **(177)** respectively; the product yield decreased in the order Et_4N^+>Li>Mg^{+2}. He concluded that the reaction temperature has effects on the product ratio and also isolated cis structure.

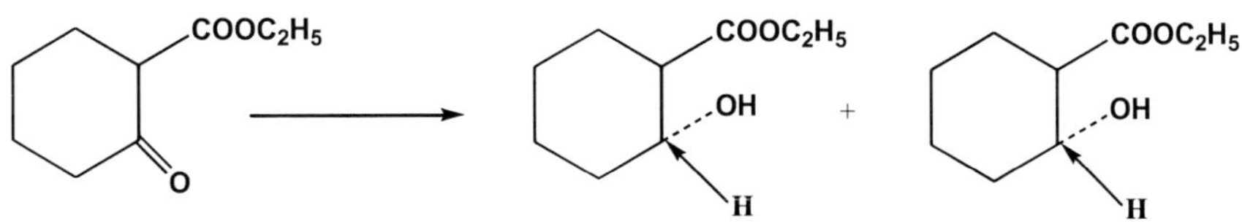

(1) **cis** **trans**

(177)

The reduction of β- ketoesters to the corresponding ethyl esters by Clmenson [110, 148] or by the Wolf – kishner method [149, 150] was not proved to be a general applicability and found that reduction of such esters by thioacetal hydrogenolysis method of Wolform and Karabines [151] can be affected under this condition. Newmann et al. [152] studied the reduction of **(1)** to hexahydrobenzoate **(178)**.

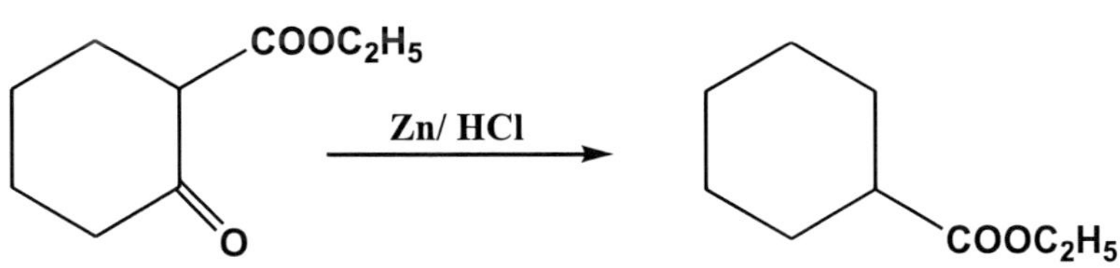

(1) **(178)**

Reduction of 2-allyl-2-carboalkoxycyclohexanones, 2-propyl-2-carboethoxycyclohexanone and 2-benzyl-2-carboethoxycyclohexanone with boron hydrides in the presence and absence of several chelating agents were studied [153, 154].

The reduction of four- to eight-membered cyclic β-keto esters to β-hydroxy esters by various yeasts and fungi is described. The microbial reduction of Et cyclobutanone-2-carboxylate is reported for the first time. The different stereospecificities observed in these reductions involving a dynamic kinetic resolution are analyzed. The enantiopure reduction products have been used in the synthesis of both enantiomers of functionalized lactones and epoxides [155].

The microbial reduction of β-keto esters derived from 2-tetralone has been shown to produce good yields of 2-hydroxy-1-carboxy- or 3-hydroxy-2-carboxy esters [156[.

A novel and efficient reduction of various prochiral ketones such as acetopehones, α-azido aryl ketones, β-keto esters, and aliphatic acyclic and cyclic ketones to the corresponding optically active secondary alcohols with moderate to excellent chemical yield was achieved by using Daucus carota, root plant cells under extremely mild and environmentally benign conditions in aqeous medium, has been described [157].

Cyclic or acyclic α-alkyl-α-hydroxy-β-keto esters **(179- 182)** were reduced by baker's yeast with high stereospecificity. For the cyclic esters, one enantiomer of the racemic mixture was transformed into optically active trans-dihydroxy compounds, while the remaining α-hydroxy-β-keto ester was enantiopure [158].

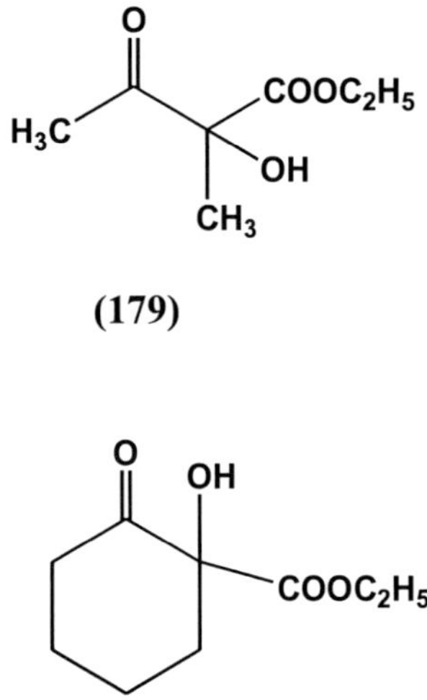

(179)

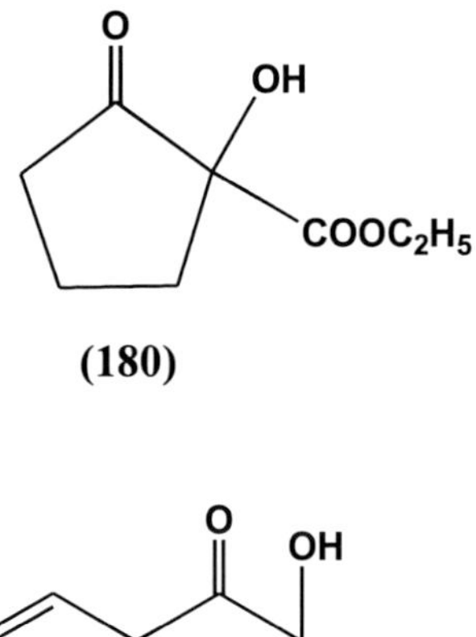

(180)

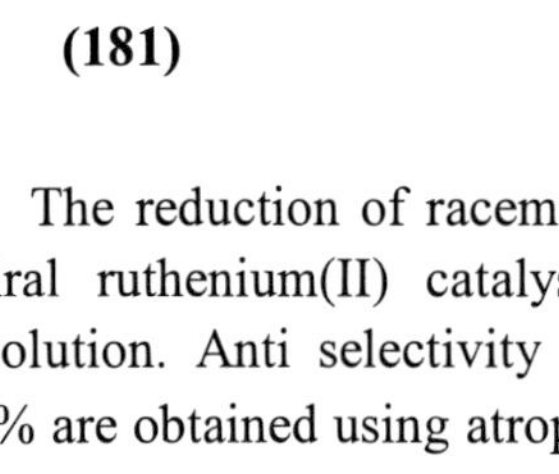

(181)

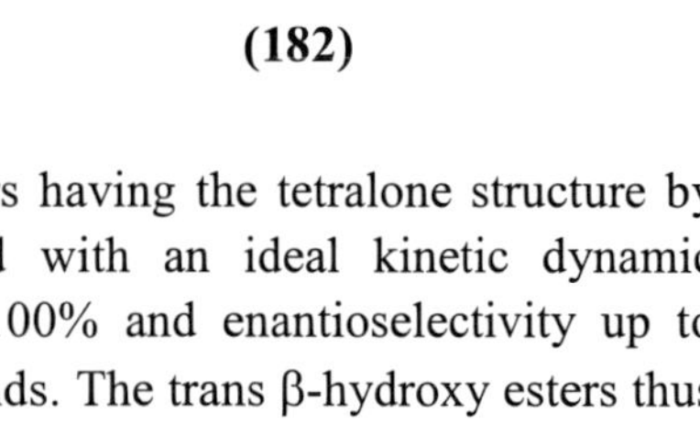

(182)

The reduction of racemic β-keto esters having the tetralone structure by chiral ruthenium(II) catalysts is realized with an ideal kinetic dynamic resolution. Anti selectivity approaching 100% and enantioselectivity up to 97% are obtained using atropisomeric ligands. The trans β-hydroxy esters thus available are useful starting materials for production of enantiomerically pure compounds [159a].

Dehydrogenases cloned from bakers' yeast (Saccharomyces cerevisiae) as fusion proteins with glutathione (S)-transferase were tested for their ability to reduce three homologous cyclic β-keto esters. The majority of dehydrogenases reduced ethyl 2-oxo-cyclopentanecarboxylate, yielding a pair of diastereomeric alcohols with consistent (1R)-stereochemistry. Ethyl 2-oxo-cyclohexanecarboxylate reductions afforded only cis-alcohol enantiomers. Ethyl 2-oxo-cycloheptanecarboxylate was accepted by two enzymes in the collection, and both yielded mainly the cis-(1R,2S)-alcohol. Escherichia coli cells overexpressing the YDL124w gene were used in a dynamic kinetic resolution of ethyl 2-oxo-cyclohexanecarboxylate to produce the key intermediate in a chemo-enzymatic synthesis of (1R,2S)-2-methyl-1-cyclohexanol, an important chiral building block[159b].

17. PECHMANN REACTION

In this reaction phenols condense with β-keto ester, and the reactivity of these esters differ very widely [160], one giving rise to coumarin, and the other to chromone .

It has been stated in the literatur that coumarin was prepared by the reaction of **(1)** [128, 161-164] with phenol in the presence of dehyderating agent like concentrated sulphuric acid [165] or phosphrous oxychloride [166].

Also, 3:4 tetrahydrobenzo-7-hydroxylcoumarin **(183)** and 3:4-tetrahydrobenzo-7- methylcoumarin **(184)** were prepared by the condensation of (**1**) with resorcinol [167] and m-cresol respectively [161].

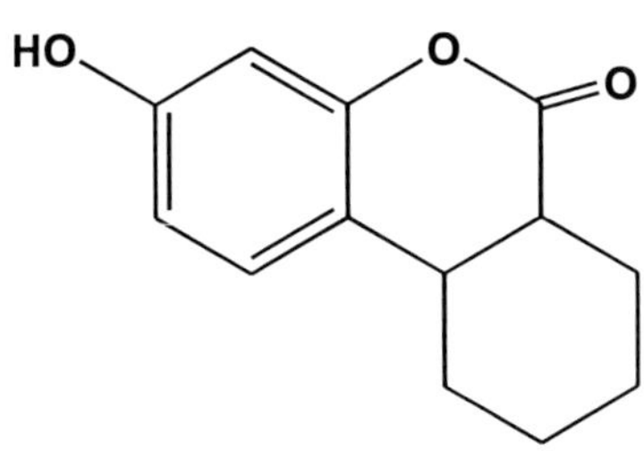

(183)

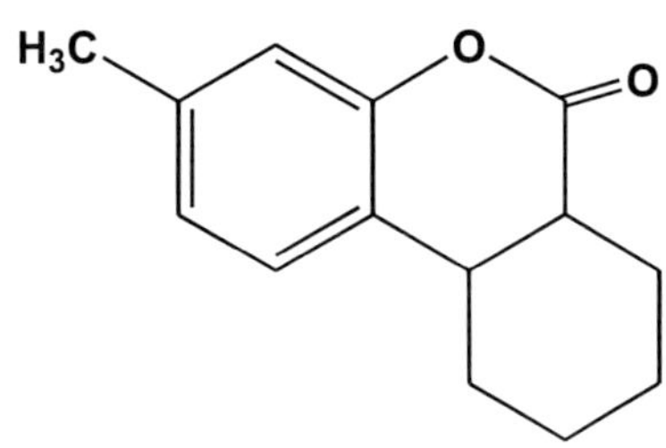

(184)

Besides, coumarin was prepared via condensation of (**1**) with 5- hydroxy-1,4-benzodioxan [168].

In fact, the biological activity of linear furocoumarins commonly known psoralenes has been known [168]; and also the synthesis of novel psoralenes [169-172] structurally related to methoxsalen **(185)** and trioxsalen **(186)** was descrided [173- 175]. Methoxsalen **(185)** was prepared via condensation of **(1)** with resorcinol in 9 steps [165, 176], while trioxsale **(186)** via condensation of **(1)** with methyl resorcinol in 2 routes involving 4-5 steps [177]. Also the condensation of **(1)** with 2-bromoresorcinol to form psoralene was stated [178].

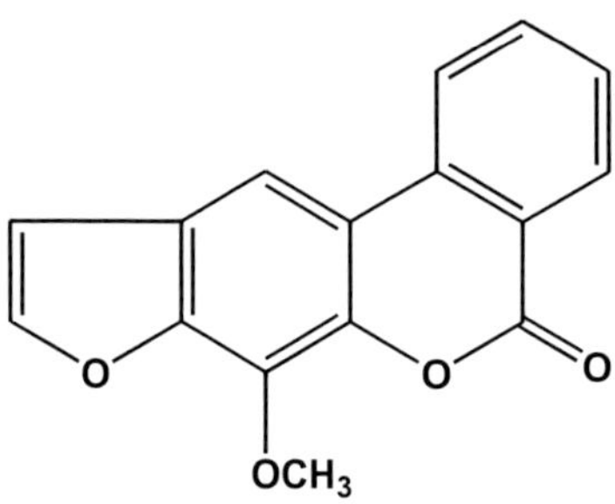

(185)

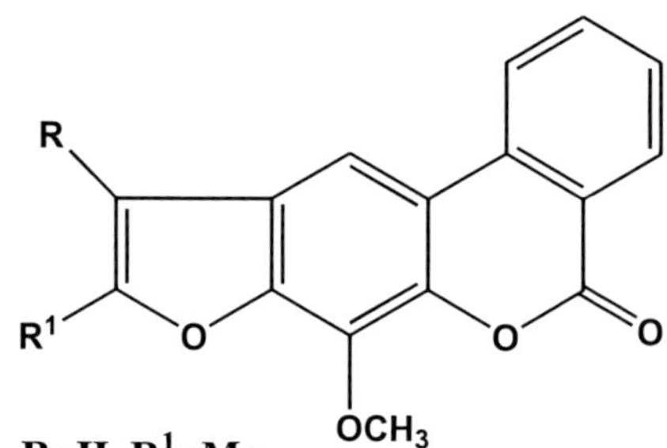

R=H, R^1=Me
R=Ph, Me,R^1=H

(186)

On the other hand, Kappe-Mayer [179] described a new modification of the Von Pechamann reaction via condensation of **(1)** with hydroxylated coumarin and quinoline derivative to give fluorecent **(187)** [180].

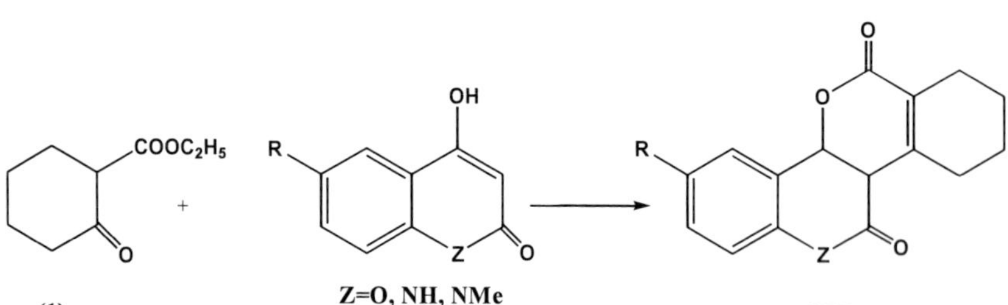

(1)

Z=O, NH, NMe
R=H, Me

(187)

Furthermore, Merchant et al. [181] utilized 4-hydroxy coumarins for the condenastion with **(1)** to produce benzopyranopyrandione **(188)** which underwent dehydrogenation to benzopyranobenzopyrandione **(189)**.

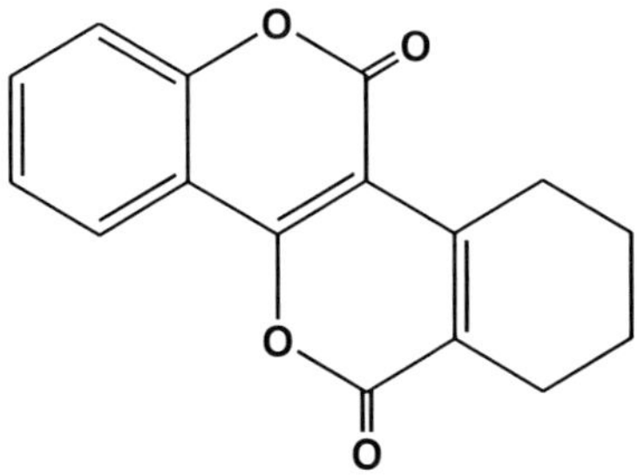

(188)

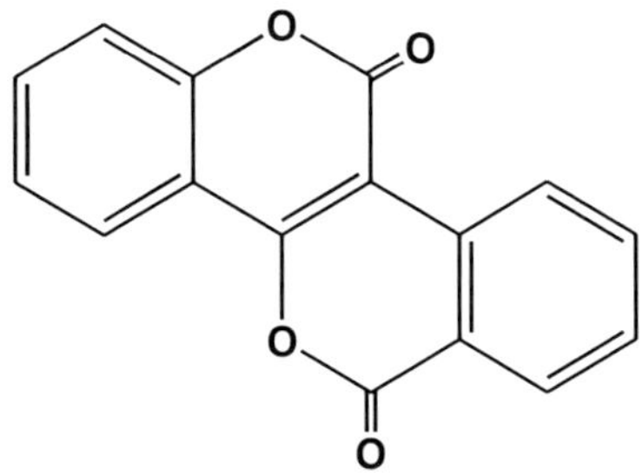

(189)

In the same manner, dibenzocoumarins **(190)** were prepared by condensation of **(1)** with naphthols [182] which on methyation, cleaved by the base and second methylation produced **(191)**.

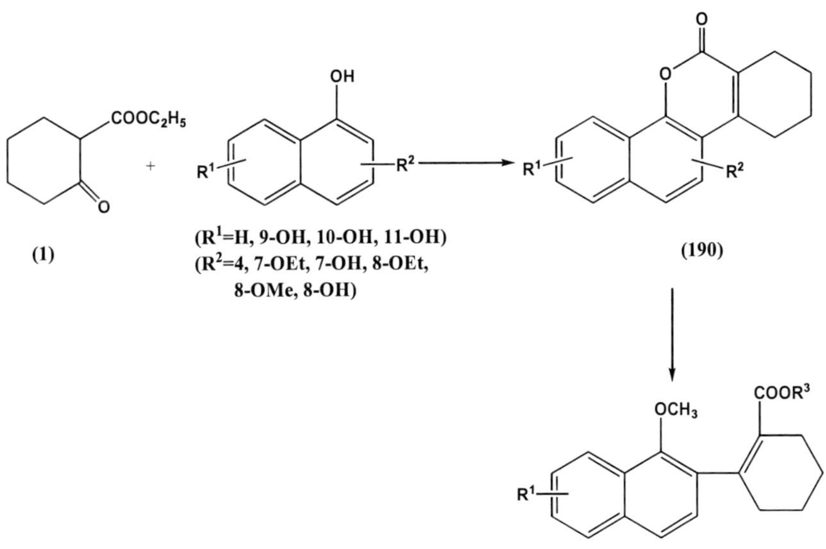

(1)

(R^1=H, 9-OH, 10-OH, 11-OH)
(R^2=4, 7-OEt, 7-OH, 8-OEt, 8-OMe, 8-OH)

(190)

Same R;R^3=H, Me

(191)

Also, antidepressant dibenzopyran derivative **(192)** was prepared via condensation of **(1)** with 4-dimethylaminophenol [183].

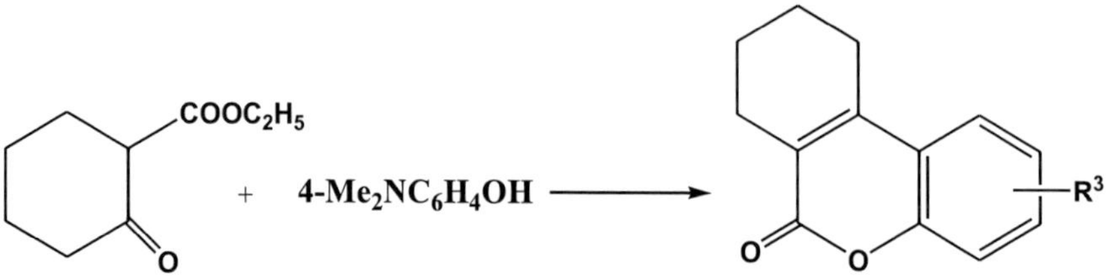

R^3=2-NMe$_2$

(1) **(192)**

It has been stated in the literature that, β-keto ester condenses with di- or polyhydric phenols in absence of dehydrating agent forming chromones [184]. Thus 2,3-cyclohexeno-7-hydroxy chromone [185] was prepared by heating **(1)** with resorcinol.

On the other hand, the synthesis of linear or angular type of furanoxanthones have been reported by many workers [186- 190[. The angular furano compound is simple, while a linear isomer needs blocking group [191]. Thus **(1)** was thermally condensed with 2-methyl resorcionl to give xanthone **(193)** which underwent severl steps to produce furoxanthone **(194)**. Also fuorxantthone **(195)** was prepared from **(1)** and resorcionl [192].

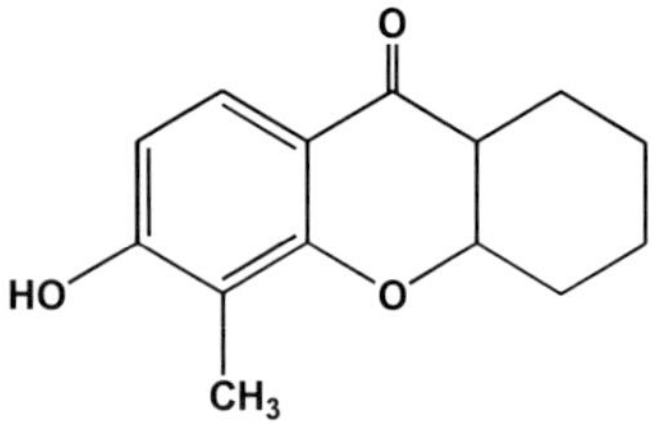

(193)

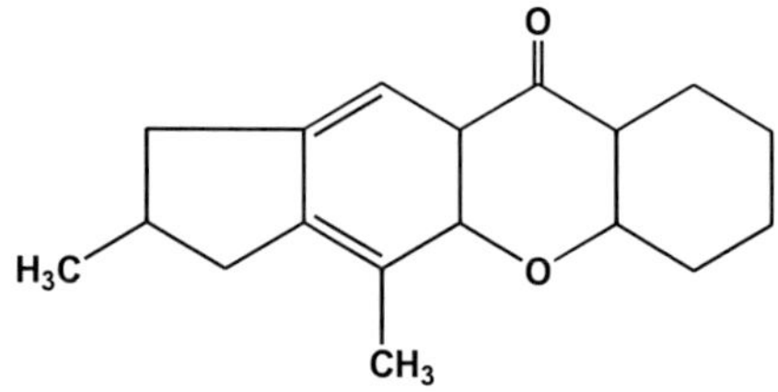

(194)

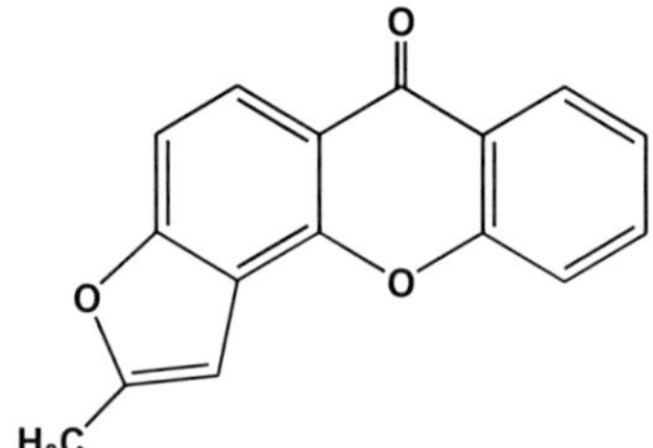

(195)

Untze [193] descrided the condensation of **(1)** with polyhdric phenols to give xanthone, chromone and coumarin derivatives **(196)** and **(197)** which were used as antioxidants.

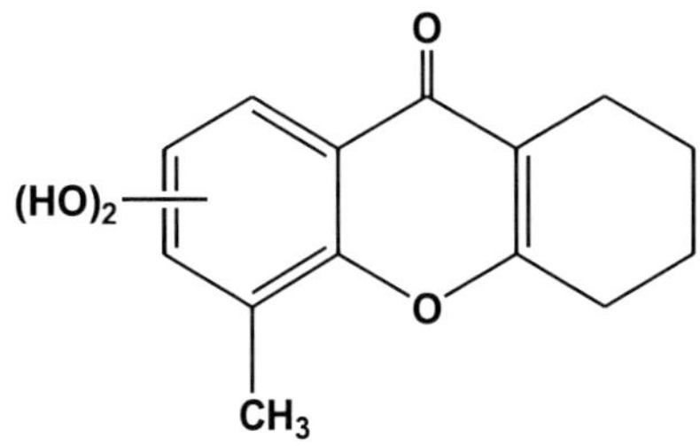

(196)

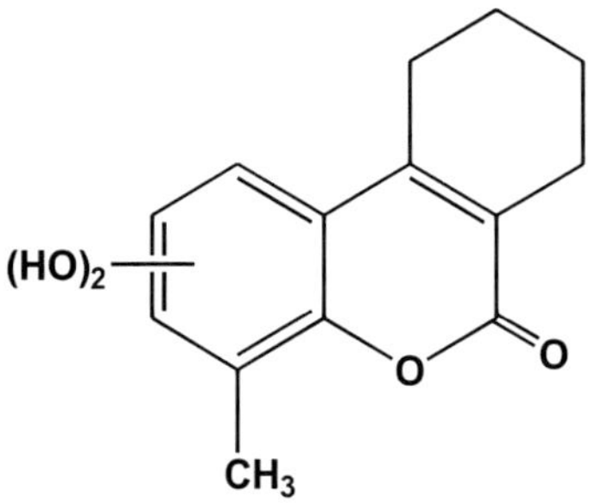

(197)

Chattergea et al. [194] described the synthesis of racemic o-tetramethylbrazilin by reduction of anhydro o-trimethylbrazilin **(198)** followed by oxidation to give cis-O-trimethylbrazilin **(199)** which underwent methylation to **(200)**.

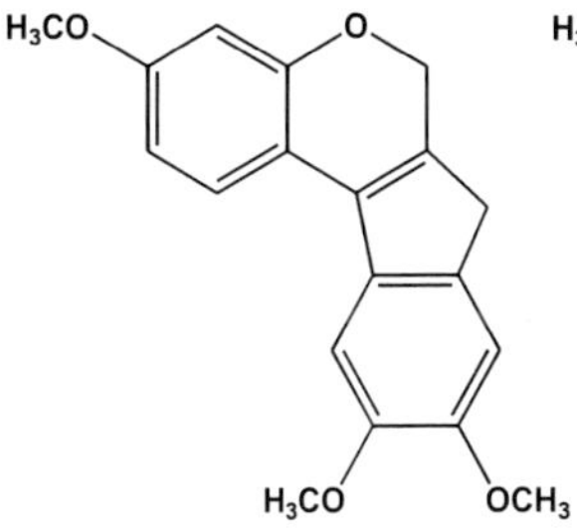

(198)

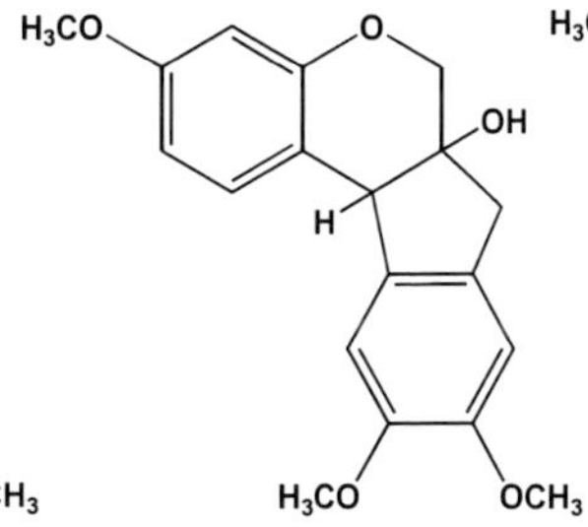

(199)

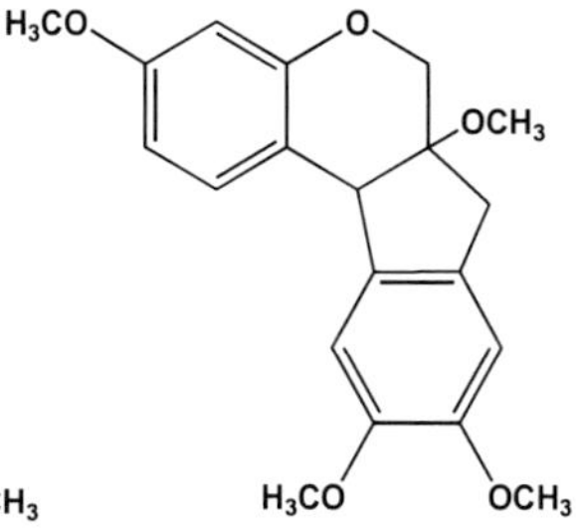

(200)

Bakre et al. [195] reported that **(1)** condensed with thiocresols (except p-thiocresol) to produce 2,3-cyclohexeno-1-thiochromone derivative **(201)**.

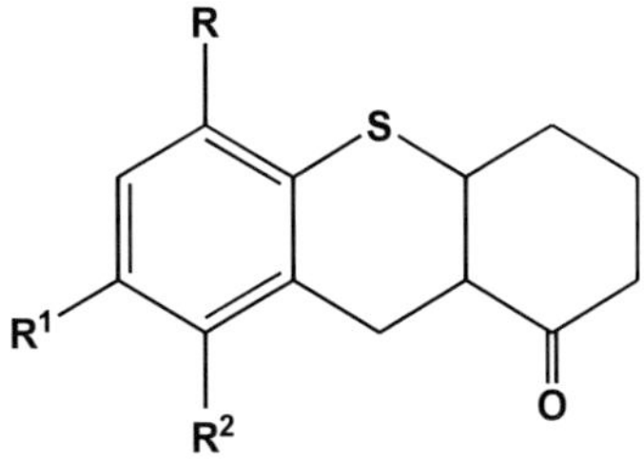

(R, R^1, R^2= H, Me)

(201)

Metwally et al. [196] reported that, the reaction of **(37)** with m-cresol and resorcinol in presence of polyphosphoric acid gave 6a,7,8,10a-tetrahydro-3,9-dimethyl (or 3-hydroxy-9-methyl)-6-oxo-7-pheny-6H-dibenzo [b,d] pyran-8-carboxylate **(202a,b)**.

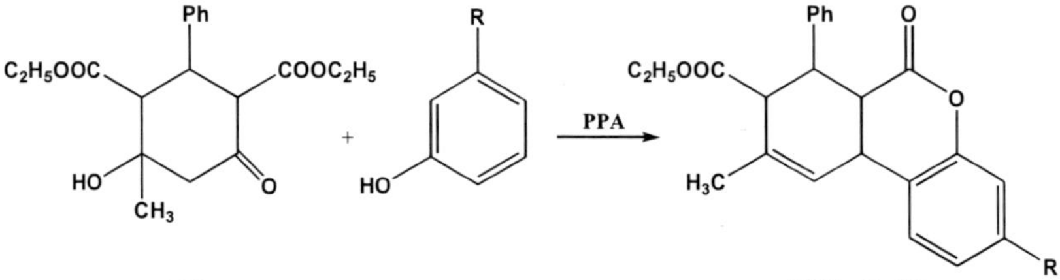

(37)

(202:a,R=Me b,R=OH)

The Pechmann reaction of **(7)** with phenols in presence of PPA has been reported to give 7,14 dihydro-3,10-di (substituted)benzo [1,2- c;4,5c] bis [1] benzopyran – 6,13- diones **(203)** [197].

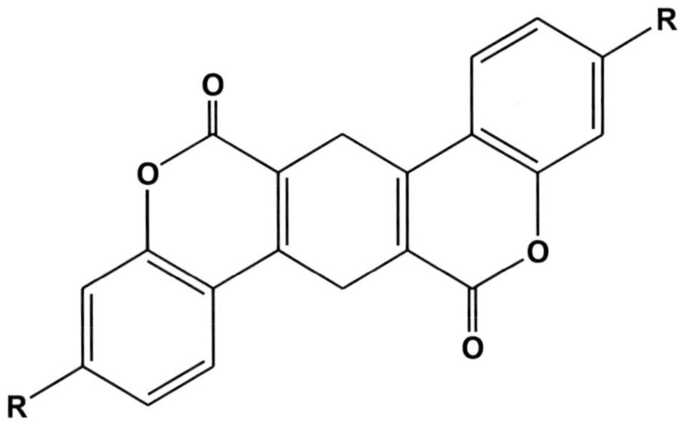

(203)

Treatment of various acyclic, cyclic, aromatic, heterocyclic β-keto esters like ethyl acetoacetate, methyl salicylate, ethyl cyclopentanone-2-carboxylate, and ethyl 2,3-dihydro-3-oxobenzofuran-2-carboxylate with 3H-7,8-dihydro-3,8-dioxo-10-hydroxypyrano[2,3-f]quinoline undergoes Pechmann condensation to yield 3H,11H-7,8-dihydro-9-methyl-3,8,11-trioxodipyrano [2,3-f; 2,3-c]quinolines, 3H-7,8,9,10,11,12-hexahydro-3,8,12-trioxopyrano [2,3-f]quinolino[3,4-b]pyrano[3,4-a]cyclopentanes, 3H,13H-7,8-dihydro-3,8,13-trioxopyrano[2,3-f]quinolino[3,4-b]-[1]-benzopyrans, and 3H,14H-7,8-

dihydro-3,8,14-trioxopyrano[2,3-f]quinolino[3,4-b]pyrano[3,4-b]-[1]-benzofurans, respectivly [198].

18. Reaction with Halogens

Reaction of 1,4-cyclohexanedione –2,5-dicarboxylates (**7**, R = Bu, Et) with bromine in carbon tetrachloride afforded 2,5-dihalo-1,4- benzoquinone-3,6-dicarboxylates **(204)** [199].

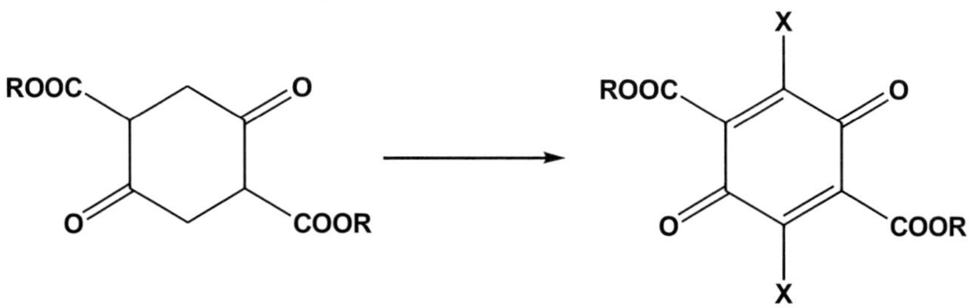

(R=Bu, Et) **(X=Br, Cl)**

(7) **(204)**

It has been reported that **(1)** undergoes selective bromination with N-bromosuccinimde in cabon tetrachloride to give 1- bromo drivative **(205)** [200] which underwent partial rearrangement to the 3- bromo ester **(206)** [201] and dehydro bromination with aniline ether or collidine to the saturated ester **(1)** [202, 203]. Also, Bernnar concluded the formation of **(207)** as illustrated in the following scheme[204, 205].

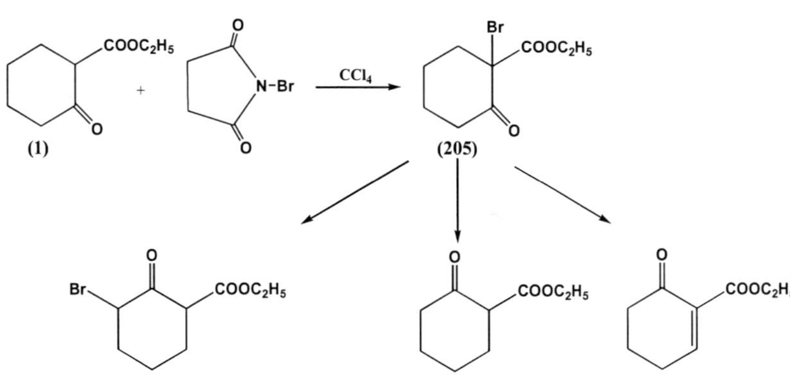

On the other hand, ethy-1-chloro-2- oxocylohexene carboxylate **(208)** was obtained via chlorination of **(1)** [206a] with sulfuryl chloride or as described by Bernner [205], which underwent dehydrochlorination with collidine to cyclohex-2-enone **(209)**. Treatment of **(208)** with ethylene glycol porduced ethyl- 1-chloro- 2,2- ethylenedioxycylohexancarboxylate **(210)**. Favorski reaction of **(211)** to ethyl cyclopent-1-ene carboxylate **(212)** porduced trans- cyclopentan –1,2-dicarboxylic acid **(213)** and the corresponding monoester **(214)**.

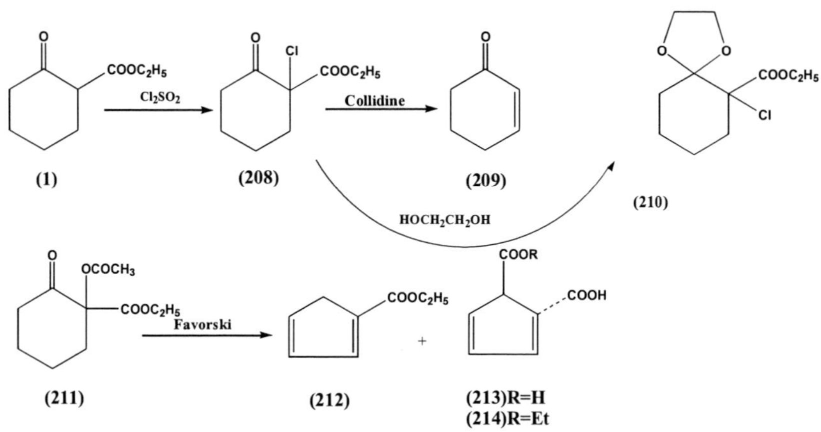

Novel chiral rare earth metal complexes bearing perfluorinated binaphthyl phosphate ligand RE[(R)-F8BNP]3 (RE = rare earth; F8BNP = 5,50,6,60,7,70,8,80-octafluoro-1,10-binaphthyl-2,20-diyl phosphate) have been synthesized and used as a catalyst for the asymmetric electrophilic fluorination reaction of β-keto esters. The use of Sc[(R)-F8BNP]3 catalyst in combination with 1-fluoropyridinium triflate (NFPY–OTf) as a fluorinating agent was found to give the desired α-fluoro-β-keto esters in high chemical yields and enantiomeric excesses (up to 88% ee) under mild conditions[206b].

19. Wittig Reaction

In the overall Wittig reaction an olefin is formed from the reaction of aldehyde or ketone and alkyl halide in which halogen bearing carbon contain at least one hydrgen in presence of phosphine. The Wittig reaction has been carried out intramolecularly to prepar rings containing from 5 to 16 C both by single ring closure [207].

According to intramolcular Wittig reaction, **(1)** was alkylated as their K-salts to give ester **(215)** which was hydrolysed and decarboxylated to give **(216)** which was treated with phoshine and cyclized to **(217)** [208].

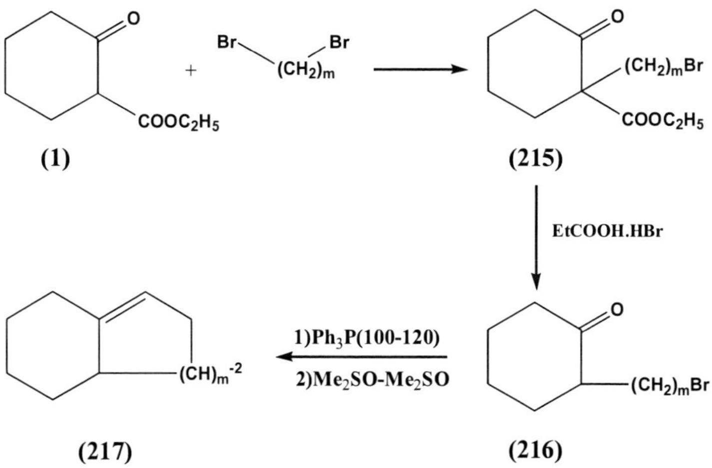

20. Decarboxylation

It has been reported that β-keto ester undergo decarboxylation by different procedures and reagents. A new method is the treament of β-keto ester with aluminium oxide in dioxan under ntirogen to give high yield of the corresponding dealkoxycarbonylation poduct and according to this procedure **(1)** gave 80% cyclohexanone **(218)** [209[.

The use of base like 1,4-diazabicyclo [2.2.2] octane is useful for the seletive cleavage of the β-keto ester to corresponding ketone [210].

On the other hand the decarboxylation of the β-keto ester by H_2O-Me_2SO in the presence or absence of added salts is a convenient prepartive rout leading to ketone [211, 212]. According to this method **(1)** undergoes facile decaboxylation on heating in Me_2SO – H_2O [213].

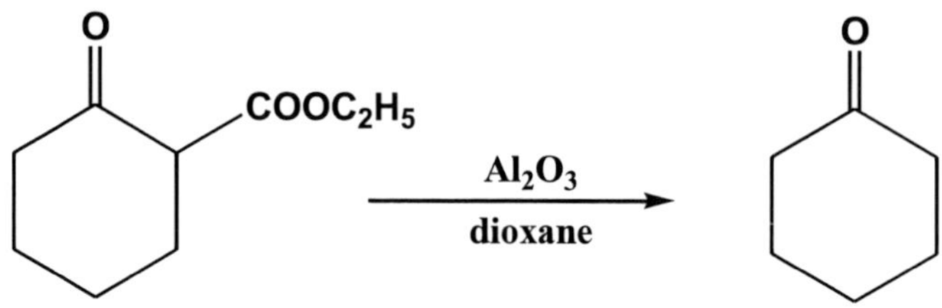

(1) (218)

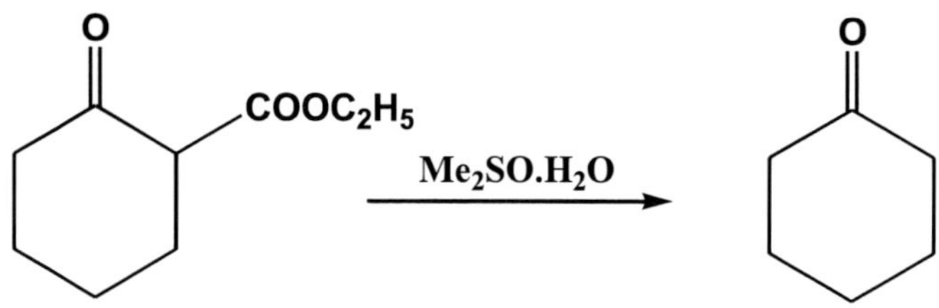

(1) (218)

Tsuda et al. [214] concluded that the halides of alkali earth metals are superior to alkali metal halides for dealkoxycarbonylation of β-keto ester. Thus heating **(1)** with $MgCl_2$ in Me_2 SO gave cyclohexanone **(218)**.

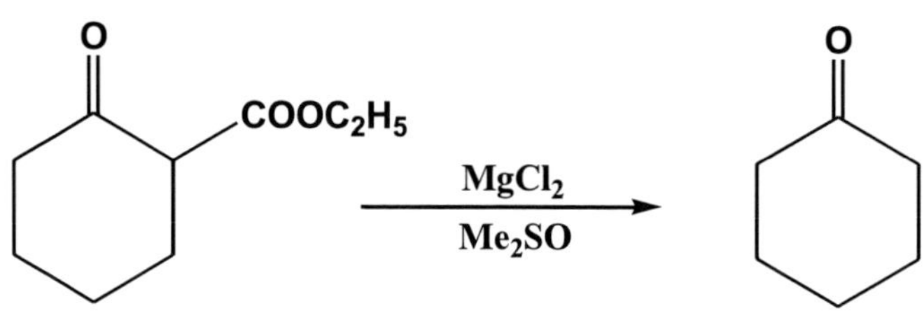

(1) (218)

21. Sulfuration

It has been reported that sulfuration of (**1**) with hydogen sulfide produced a mixtures of cyclohexenethiol **(219)** [A and/or B] thiones **(220)** as illustarted [215].

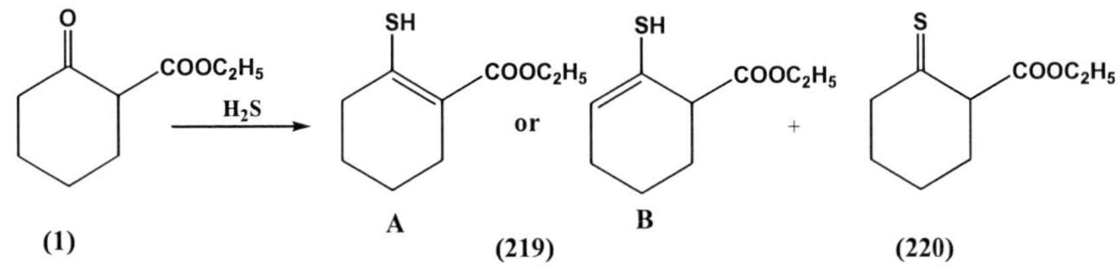

On the other hand, 1,2- dithiole –3- thione **(221)** was prepared by reaction of **(1)** with phosphororus pentasulfide in xylene [216].

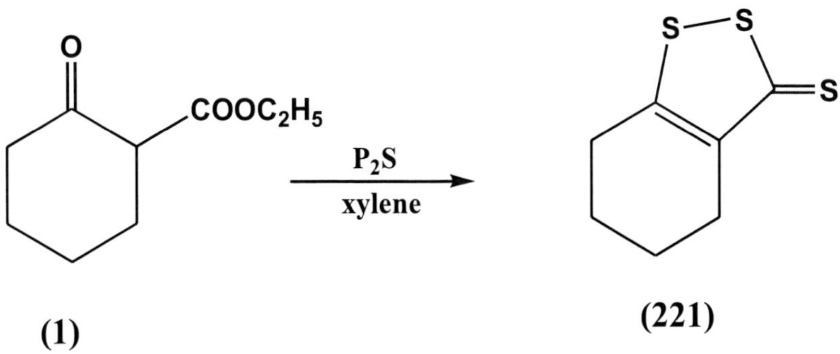

De Groot and Jansen [217] described the formation of α- sulfenylated product **(222)** by quenching the enolate of **(1)** with PhSCl or MeSCl as nuleophiles.

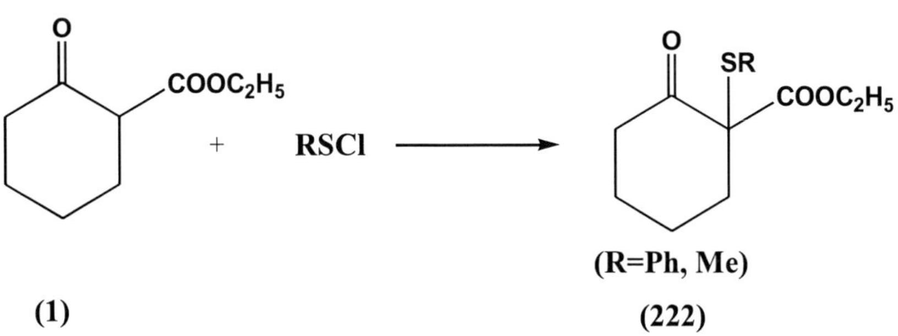

22. Knoevenagel Reaction

Knoevenagel reaction of β-keto esters [218] with malonic acid derivative is one of the most customary and useful methods for the synthesis of 2-pyridinols and 2,3-pyridinols. Also the latter have been obtianed mostly from Knoevenagel reaction of β-keto ester and cyanoacetamide and normal Knoevenagel reaction products were generally assumed as intermediates [219, 220] in these reactions.

According to Knoevenagel, **(1)** reacts with malononitrile to give alkoxypyridinol **(223)** [221] according to the following scheme:

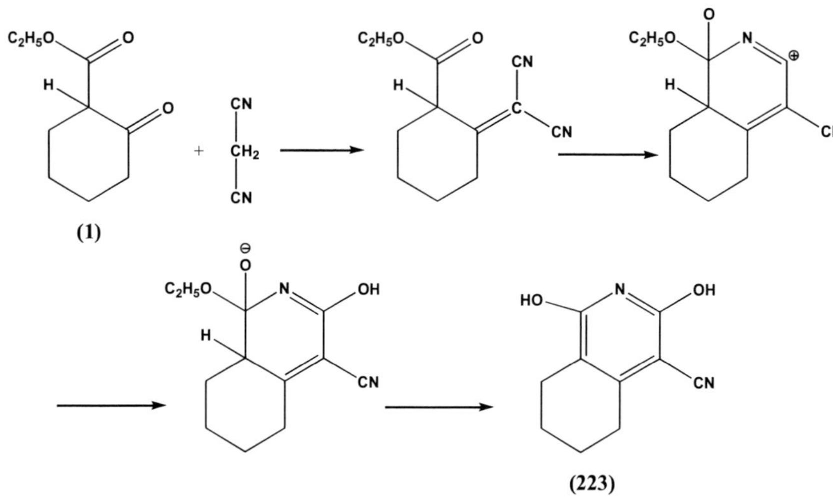

23. Ring Expansion

It has been reported that, on treating potassuim salts n- membered cyclic keto esters with pheny1-1-propenylselenone, a one carbon ring expansion reaction involving cyclopropation or vinylation took place [222]. Treatment of the potassuim salt of **(1)** with phenyl-(E) –propenyl selenone 2- (E) in THF at room temperature led to a predominant formation of bicyclo [5.1.0]octane **(224)** together with a small amount of l-vinyl-3-cycloheptanone -1-carboxylate **(225)**, whereas **(226)** was obtained as a sole product from the reaction with (Z)- porpenylselenone2- (Z)- under similar conditions .

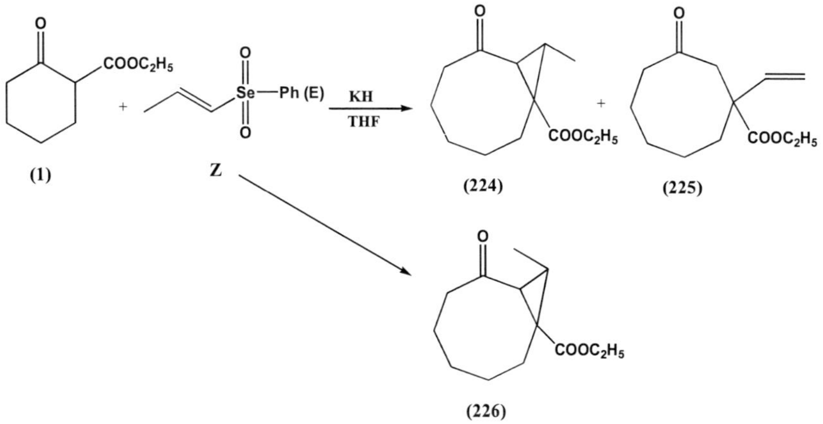

Radical ring-expansion reaction of various α-halomethyl cyclic β-keto esters and α-halomethyl benzocyclic β-keto esters and chain-extension reaction of α-halomethyl β-keto esters with Zn powder and refluxing in aqeous alcohol were carried out to generate the corresponding ring-expansion and chain-extension products [223].

Reactivities of tetraphenyldisilane (TPDS), tris (trimethylsilyl)silane, and tributyltin hydride in the radical ring expansion of α-haloalkyl cyclic β-keto esters were examined. Among these reagents, TPDS was found most effective for the preparation of medium-sized cyclic compounds in terms of yields and ring-expansion/reduction selectivity [224].

SmI2 reduction of some aromatic as well as aliphatic α-bromomethyl cyclic β-keto esters produced one-carbon homologated γ-keto esters in modest to good yields. The samarium iodide-mediated ring expansion of 2-bromomethyl-1,2,3,4-tetrahydro-1-oxo-2-naphthalenecarboxylic acid ethyl ester gave 6,7,8,9-tetrahydro-5-oxo-5H-benzocycloheptene-7-carboxylic acid ethyl ester [225, 226].

Some β-keto esters, β-diketones and enol silyl ethers were converted in 60-77% yields into the corresponding α-(oxoquinazolinyl)amino cyclic ketone derivatives by reaction with 3-(acetoxyamino)-2-(1-methylethyl)-4(3H)-quinazolinone. Further oxidation of some of these products with lead tetraacetate gives products whose nature depends on the solvent used. Ring expansion of compounds which contain 6-membered ring ketones does not occur and the only isolated product in each case is a benzoxazinone compound [227].

Free radical-promoted, C1 ring expansion of cyclic β-keto esters **(227)** (n = 8, 10, 11) gave γ-keto esters **(228)**. The method was extended to a C3 ring expansion of cyclododecanones **(229)** to (±)- and (R)-(-) muscone [228].

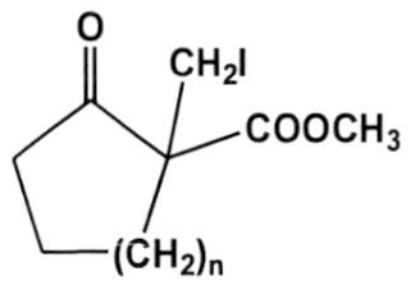

(227)

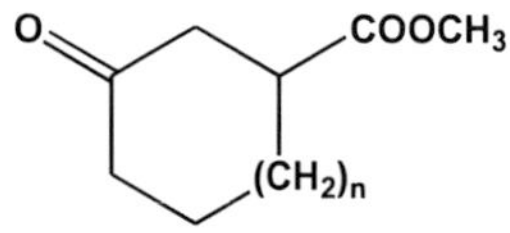

(228)

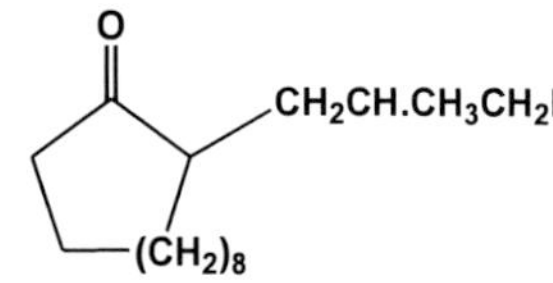

(229)

Cycloalkanones **(230)** (n = 1, 2, 3; R^1 = Et, Me; R^2 = MeCO, CO_2Et, cyano) were treated with Me_3COK in Me_2SO to give ring enlargement products **(231)**. Cyclododecanone derivative **(232)** underwent a similar expansion, and the product was converted to muscone [229].

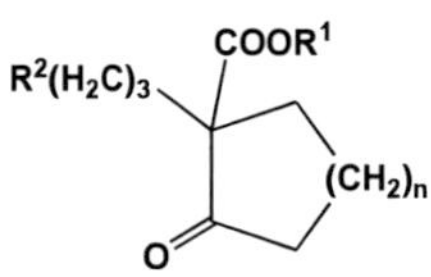

(230)

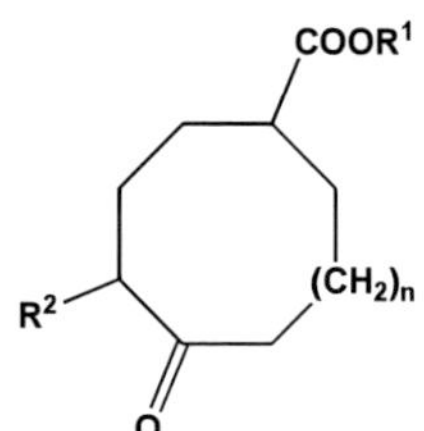

(231)

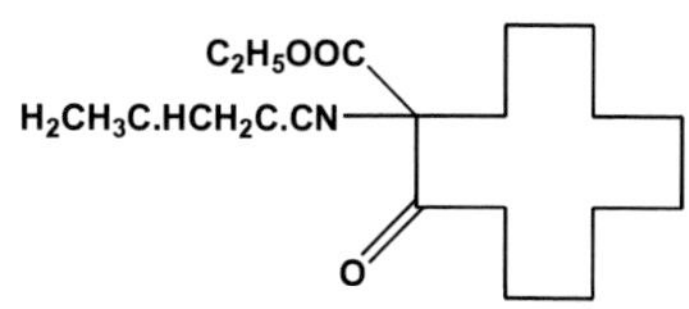

(232)

The sodium salts of 2-(methoxycarbonyl)cycloalkanones (n = 3-6, 10) reacted with $MeO_2CC{\equiv}CCO_2Me$ to yield ring-expanded 1-hydroxy-2,3,4-tris(methoxycarbonyl)cycloalkanones (same n). Fused aryl derivatives reacted similarly [230].

Previous investigation into the reaction of cyclic-β-keto esters with dimethyl acetylenedicarboxylate revealed that a ring expansion took place. X-ray crystal structure analyses were undertaken to confirm the structure of the resulting 1H-1-benzoazonine **(234)** and its precursor **(233)**[231].

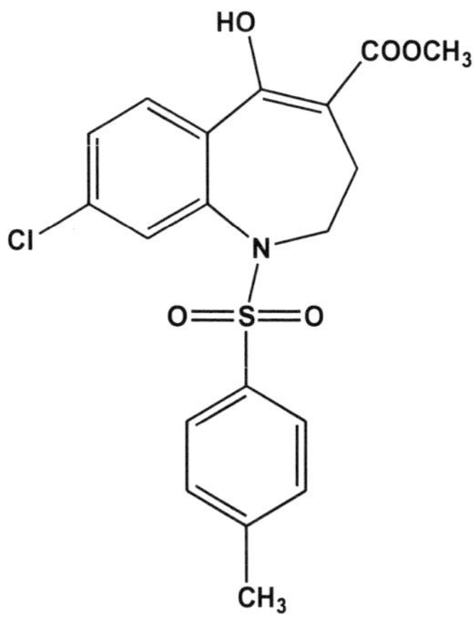

(233)

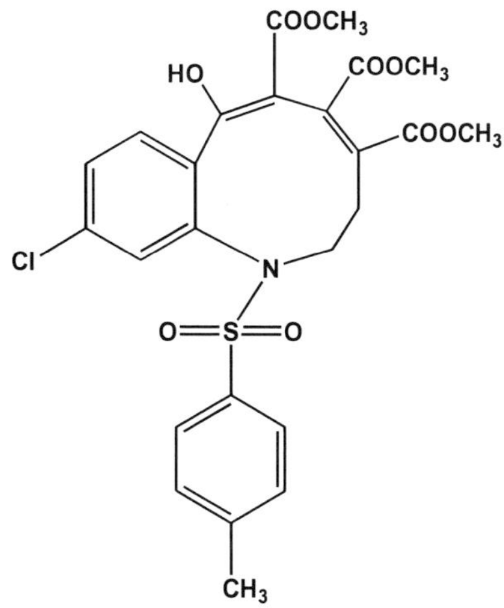

(234)

24. Schmidt Reaction

Adamson [232] applied the Schmidt reaction on ethyl cyclohexanone-2-carboxylate **(1)** and obtained lysine **(235)** in a good yield using hydrazioc acid in benzene.

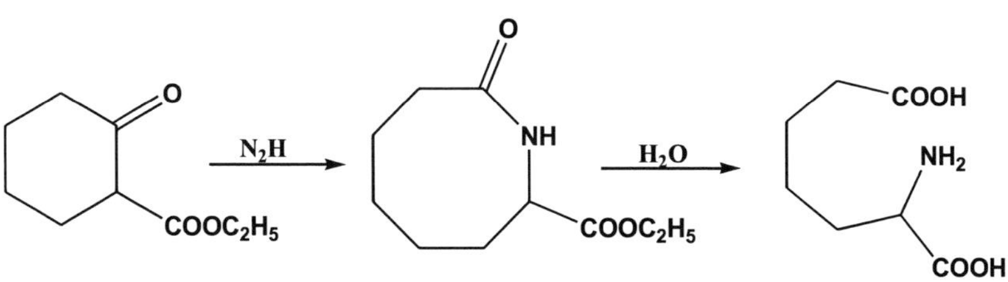

(1)

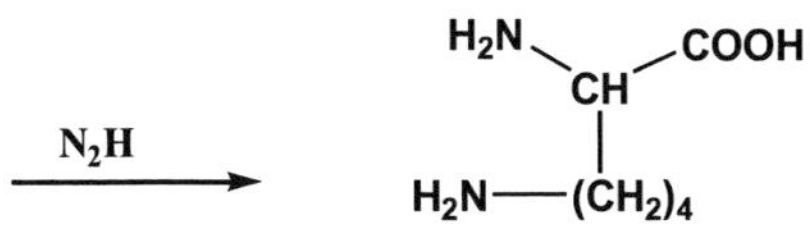

(235)

25. KETALIZATION

It has been reported that **(1)** [233] reacted with ethylene glycol in the persence of p-toluenesulphonic acid to give 1,4-dioxospiro[4,5] decan –6-carboxylate **(236)**.

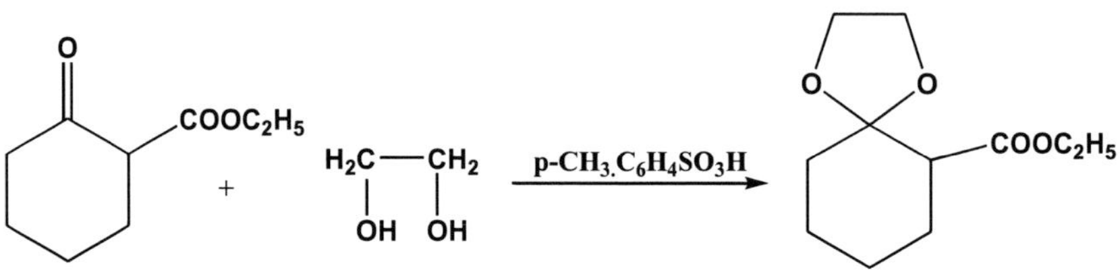

(1) **(236)**

The title compounds (**237**; R^1 = C1-10 alkyl, cycloalkyl, alkaryl, (un)substituted aryl; R^2, R^3 = C1-10 alkyl, cycloalkyl, aralkyl; n = 1-3), useful in the curing of resins, such as unsaturated polyester resins, are prepared by the acid-catalyzed ketalization of cyclic β-keto esters with aliphatic hydroperoxides. Thus, ethyl 2-oxo-1-cyclohexanecarboxylate was ketalized with tert-butyl hydroperoxide in the presence of sulfuric acid to produce ethyl 2,2-di(tert-butylperoxy)-1-cyclohexanecarboxylate (m.p. 36-38°) [234].

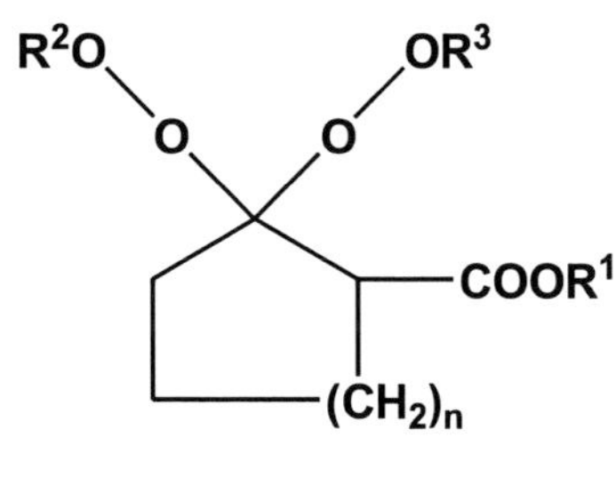

(237)

26. EFFECT OF ULTRAVIOLET LIGHT

Diethyl 1,4-cyclohexanedione-2,5-dicarboxylate **(7)** was dimerized by the action of UV light in water to give tetraethyl anil-1, 4, 4a, 4b, 5, 8, 8b-octahydro- 3, 4b, 7, 8b- tetrahydroxy- 2- 4a, 6, 8abiphenylenetetracarboxylate **(238)** [235] which is used in the manufacture of linear polyesters and polyester fibers.

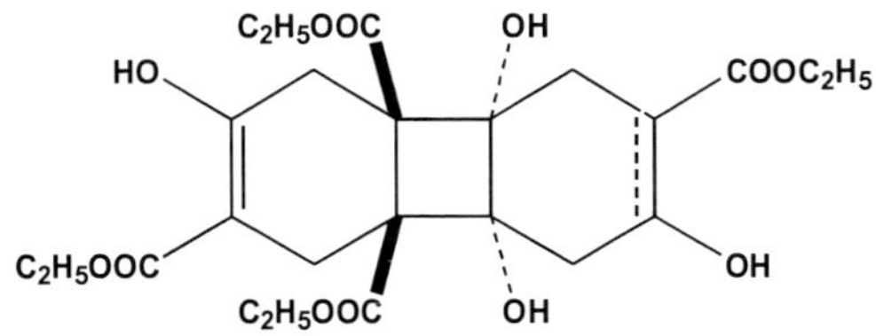

(238)

27. Hydroxylation

Non racemic 2-hydroxy-1-oxoindanecarboxylates such as **(239)** and a nonracemic 2-oxo-1-hydroxy-1-cyclopentanecarboxylate are prepared by enantioselective hydroxylation of cyclic β-keto esters with organic hydroperoxides such as cumene hydroperoxide and alkaloids such as dihydroquinine; the α-hydroxy-β-keto esters undergo stereoselective reduction with 4-ethylmorpholine-borane complex to yield trans-α,β-dihydroxy esters such as **(240)** [236].

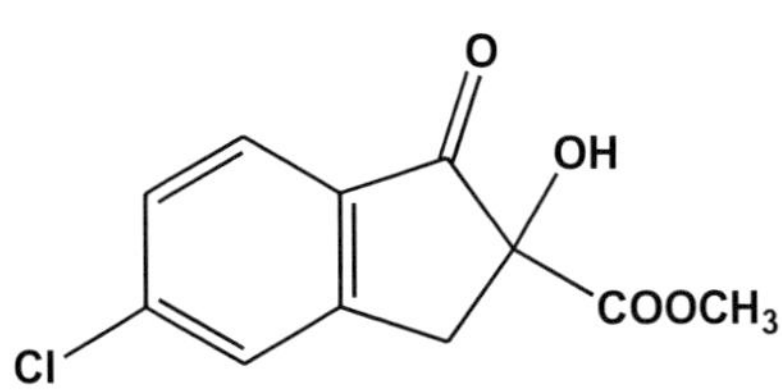

(239)

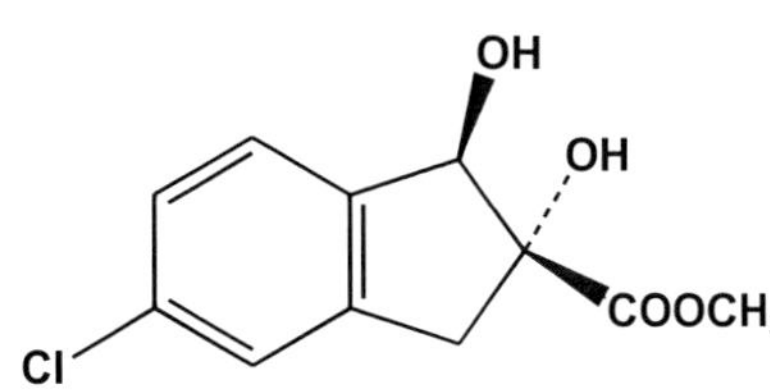

(240)

Cyclic and acyclic α-alkyl-α-hydroxy-β-keto esters, e.g. **(241)**, were obtained with 50-100% isolated yields from the corresponding β-keto esters by oxidation with moecular oxygen in a mixture of CH_3CN/i-PrOH and in the presence of cobalt(II) chloride [237].

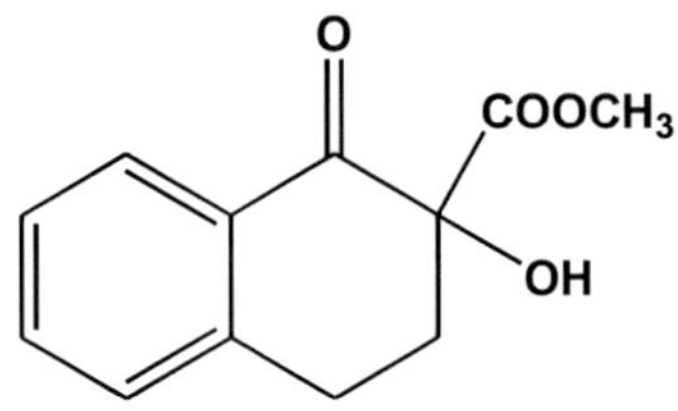

(241)

28. Bromination and Chlorination

The first general catalytic asymmetric bromination and chlorination of β-keto esters has been developed. The reactions proceed for both acyclic and cyclic β-keto esters catalyzed by chiral bisoxazolinecopper(II) complexes giving the corresponding optically active α-bromo- and α-chloro-β-keto esters in high yields and moderate to good enantioselectivities [238].

α-Bromination of carbonyl compounds (cyclic and acyclic ketones, amides, and β-keto esters) has been achieved efficiently by treatment with N-bromosuccinimide (NBS) and catalyzed by silica-supported sodium hydrogen sulfate ($NaHSO_4.SiO_2$) [239].

A simple and rapid process was developed for the α-monohalogenation of 1,3-keto-esters with N-halosuccinimides catalyzed by Amberlyst-15 at room temperature to produce the corresponding 2-halo 1,3-keto-esters in high yields. This protocol also extended to α-halogenation of cyclic ketones [240[.

An efficient catalytic enantioselective electrophilic fluorination of cyclic and acyclic β-keto esters is reported. As low as 1 mol% of chiral bis (oxazoline)-copper triflate complexes catalyzed the fluorination by means of N-fluorobenzenesulfonimide [241].

Optically active α-fluoro ketones are prepared by reaction of ketones with fluorinating agent in the presence of an asymmetric transition metal complex of formula [M_2LpAq]y-(Z-)y or MLrBsDc [wherein L = an asymmetric ligand; Z- = an anion; A = a monovalent anionic ligand; M = a transition metal; p = 2, 4; y = 0, 2; B = H_2O, a neutral ligand; D = an anionic ligand or counter anion; r = 1,2; s = an integer of 0-6; c = an integer of 0-2]. Thus, [[Pd[(R)-dtbm-segphos](μ-OH)]2]2-(OTf-)2 (**242**; Ar^2 = 3,5-di-tert-butyl-4-methoxyphenyl) was treated with 2-oxocyclopentanecarboxylic acid tert-butyl ester and N-fluorobenzenesulfonimide to give 1-fluoro-2-oxocyclopentanecarboxylic acid tert- butyl ester. These compounds are intermediates for drugs, agrochemicals, or many other versatile chemicals [242].

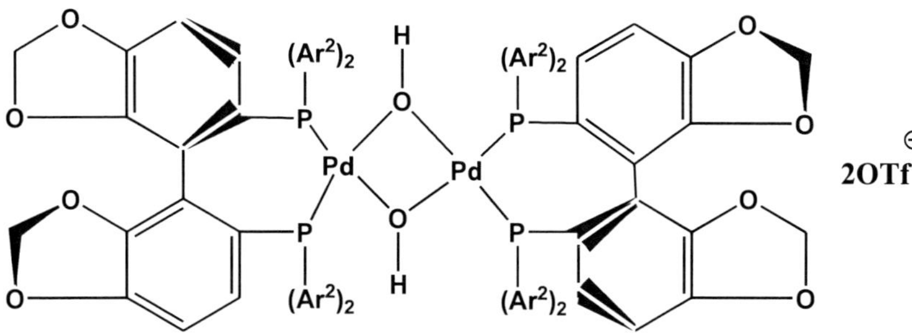

(242)

The reactions of β-fluoro vinamidinium salt $RR^1NCH:CFCH:N\text{-}RR^1$ [$RR^1 = (CH_2)_5$] with enolates derived from β-keto esters at 80°C gave the cyclization products, fluorinated isophthalates, in good yields [243].

29. Transesterification

Transesterification of β-keto esters with a variety of alcohols to the corresponding β-keto esters has been carried out efficiently over aluminosilicates simply by refluxing in toluene under liquid-phase conditions. Aliphatic β-keto esters exhibit the higher reactivity than the aromatic and/or cyclic β-keto esters, whereas other esters such as α-keto esters, α,β-unsaturated esters, normal esters, and α-halo esters fail to undergo the transesterification [244].

30. Annelation Reaction

The effect of γ-substituents on the bicycloannelation of cyclic β-keto esters in presence of Pd and 1,3-allylic diacetate depends strongly on the β-ketoester structure and the bicyclo-annelation reaction has its limitations [245, 246].

A new annelation reaction in a series of cyclic Schiff's bases is described with examples of [3 + 3] cyclocondensation of 1-alkyl-3,4-dihydroquinolines and acetoacetic esters [247].

31. Cyclization with 5-Aminopyrazoles

The reaction of 4-(arylazo)-2,5-diaminopyrazoles with cyclic β-keto esters or acetoacetanilides afforded pyrazolo[1,5-a]pyrimidine derivatives **(243)** [248].

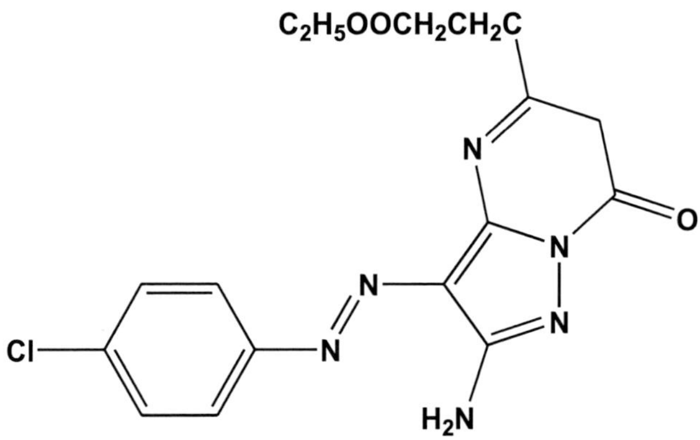

(243)

32. Coupling With α-(N-Carbamoyl) Alkylcuprates

α-(N-Carbamoylalkyl)cuprates, generated in situ from N-Boc protected dialkylamines $Me_3COCONR^1R^2$ [$R^1 = R^2 = Me$; $R^1R^2 = (CH_2)_4$, $(CH_2)_5$, $(CH_2)_6$], couple with enol triflates derived from carbocyclic and heterocyclic β-keto esters $R^3COCHR^4CO_2Et$ [$R^3R^4 = (CH_2)_4$, $(CH_2)_2N(CH_2Ph)CH_2$, etc.] to give the corresponding γ-carbamoyl-α,β-enoates in good to excellent yields. The γ-amino-α,β-enoates, available via N-Boc deprotection, can be cyclized to annulated pyrrolin-2-ones [249].

33. Tautomerization

NMR and kinetics studies of the tautomeric equilibrium of 2-acetylcycloalkanones, 2-acetyl-1,3-cycloalkanediones, and Me 2-oxocycloalkanecarboxylates of five- and six-membered rings have been carried out by comparing 1H NMR and ^{13}C NMR spectra recorded in D_2O-$DOCD_3$ and $CDCl_3$, with those results derived using a bromination kinetic procedure in diluted aqueous solution [250], where as the ring size increased, the tautomeric equil. shifted toward the enamine form[251[.

34. Aldol And Dehydration

Cyclic β-keto esters and β-keto amides undergo, in a one-pot process, an unprecedented DBU-MeOH-promoted regio- and stereoselective γ-functionalization with aldehydes, by a directed γ-Aldol reaction and dehydration sequence, to afford synthetically valuable alkylidene (or arylidene) cycloalkanones in good yields [252],.

35. Reaction with Acetylene

The enolates of cyclic β-keto esters react with electrophilic acetylenes to give the corresponding Michael adducts in good yields when the reaction is performed in acetone in the presence of catalytic amounts of K_2CO_3. The Michael adducts resulting from ethynyl methyl ketone, when refluxed in toluene in the presence of catalytic amounts of TsOH, undergo an intramolecular Aldol reaction leading mainly to bicyclo[n.3]alkadienones besides Robinson annulation products [253].

36. Biotransformation

Algal transformations of 3 cyclic β-keto esters, methyl 2-oxocyclopentanecarboxylate (MCPC), ethyl 2-oxocyclohexanecarboxylate (ECHxC), and methyl 2-oxocycloheptanecarboxylate (MCHpC), were photoautotrophically studied with an axenic strain of C. pyrenoidosa Chick. The 3 cyclic β-keto esters were transformed in 2 manners; reduction of the carbonyl group to a hydroxy group and elimination of the alkoxycarbonyl group [254].

37. Reaction with 1,4- Dihalobutenes:

Cyclic β-keto ester monoanions react with 1,4-dihalobutenes to give C-alkylated products which subsequently undergo a stereoselective SN2' O-alkylation reaction to yield functionalized enol ethers[255].

38. Cyclization of Alkenyl β-Keto Esters and Amides

In the presence of catalytic amounts of $PdCl_2(MeCN)_2$ and $Ln(OTf)_3$, various alkenyl β-keto esters and amides underwent regioselective cyclization reactions to give six-, seven-, or eight-membered-ring carbocycles in moderate to excellent yields [256].

39. Hydrogenation

The stereo controlled synthesis of a C15-C24 fragment of dolabelides is reported. The C19 and C21 hydroxyl-bearing stereocenters were installed using ruthenium-mediated asymmetric hydrogenations of a cyclic hemiketal and a β-keto ester [257, 258].

The hydrogenations of methyl 2-oxocyclopentanecarboxylate, ethyl 2-oxocyclohexanecarboxylate and 2-methylcyclohexanone on unmodified Raney Ni catalyst lead predominantly to the formation of the cis hydroxy diastereoisomers [259].

40. Reaction with Serine-Derived Cyclic Sulfamidates

Ring-opening of N-(PhF)serine-derived cyclic sulfamidate **(244)** was achieved with different nucleophiles (β-keto esters, β-keto ketones, di-Me malonate, nitroethane, sodium azide, imidazole, and potassium thiocyanate) to prepare a variety of amino acid analogs [260].

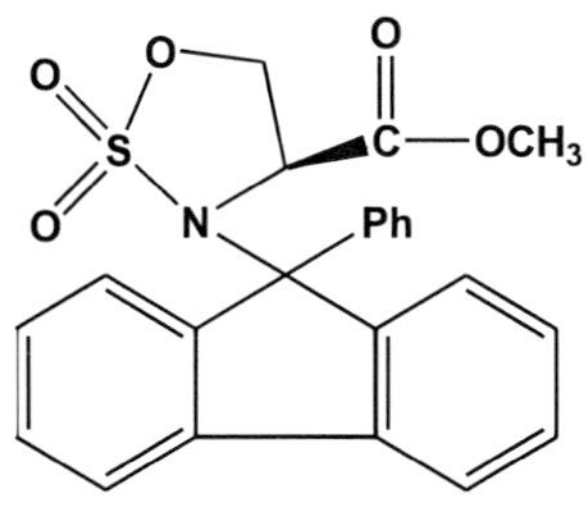

(244)

41. Ritter Reaction

Indanol intermediates **(245)** (R = H, Me, Ph, 2-thienyl), prepared via Michael addition of 1-indanone β-ketoester and acrylonitrile followed by reduction or Grignard reaction of the ketone group, were submitted to intramolecular Ritter reaction using various acid reaction conditions to produce tricyclic lactams **(246)** [261].

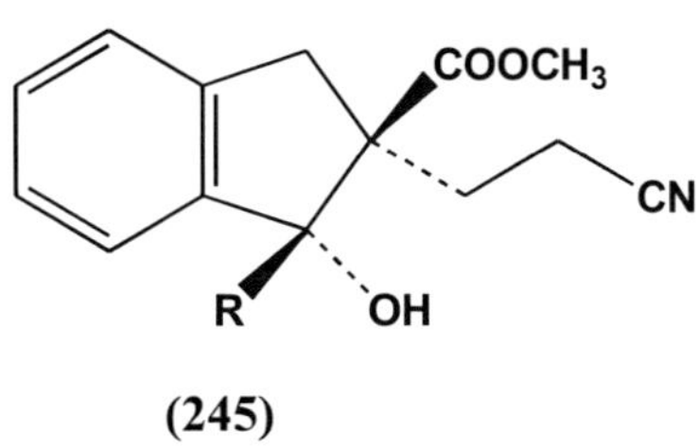

(245)

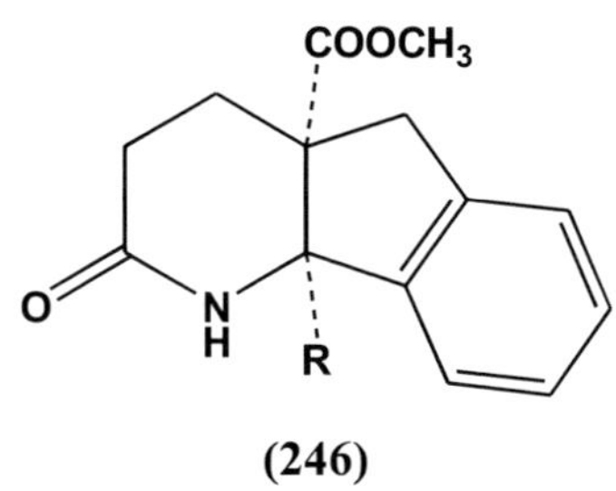

(246)

42. Sulfenylation

The α-sulfenylation of carbonyl compounds with 3-(phenylthio)-2-(N-cyanoimino)thiazolidine proceeds smoothly. Furthermore, the optically active 4-diphenylmethyl derivative was synthesized as an asymmetric sulfenylation reagent, which realized 96% ee upon α-sulfenylation of a cyclic β-keto ester [262].

Quenching the enolates of cyclic β-diketones or 2-oxocycloalkanecarboxylates with PhSCl or MeSCl produces the corresponding α-sulfenylated products [263].

43. Reaction with Sodium Hydride

Cyclic and acyclic β-keto esters, a typical cyclic α-diketone and a typical cyclic β-diketone were converted to the corresponding vinyl nonaflates by reaction with NaH in DMF at 20-55°C followed by treatment with perfluoro-1-butanesulfonyl fluoride at 20°C [264].

44. Reaction with Chiral Cyclic Diols

Chiral tricyclic α,β-unsaturated lactones such as **(247)** were easily synthesized from chiral cyclic diols and cyclic β-keto ester **(1)** [265].

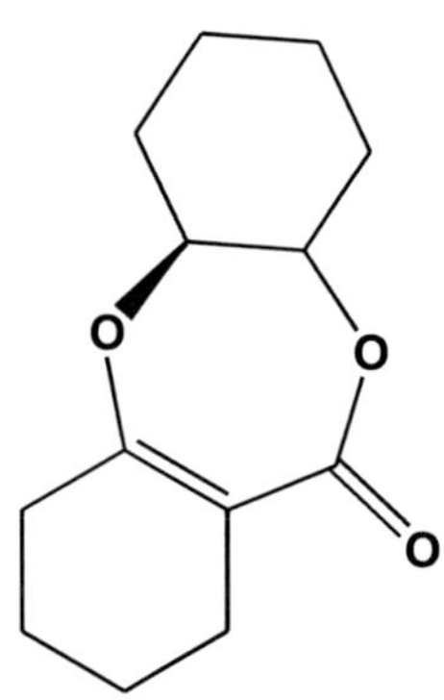

(247)

45. Dehydrogenation

Cyclic β-keto esters and β-keto amides are converted to 2-(alkoxycarbonyl)- and 2-(aminocarbonyl)-2-cycloalken-1-ones by treatment with $Pb(OAc)_4$ and $Cu(OAc)_2$ in benzene. Thus, Me 2-oxocyclopentanecarboxylate **(55)** was treated with pyrrolidine and 4-

dimethylaminopyridine in PhMe and then with $Pb(OAc)_4$ and a catalytic amount of $Cu(OAc)_2$ in C_6H_6 to give 67% of the unsaturated oxo amide **(248)** [266].

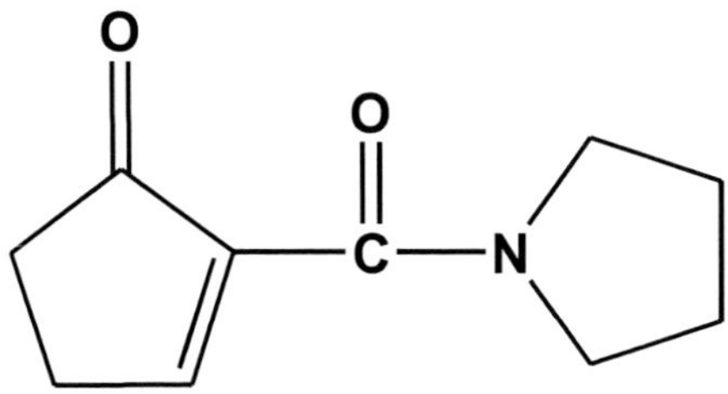

(248)

46. Regioselective Conversion

Cyclic β-keto esters, available by regioselective acylation of cycloalkanone enolates, are rapidly transformed to α,β-unsaturated acids. This functionality transposition allows the derived 3-hydroxy-4-methylthiazole-2(3H)-thione derivatives to serve as precursors to synthetically useful vinylic bromides [267].

47. Reaction With Furan Derivatives

2-(2-Furyl)cycloalkanones were synthesized by condensation of 2,5-dimethoxy-2,5-dihydrofuran with cyclic β-keto esters in an acidic medium. The decarbalkoxylation is then realized in the presence of lithium chloride and in a polar aprotic solvent, according a modified Krapcho reaction. This synthesis is extended to cyclopentanone, cyclohexanone, tetralones (simple, 6-methoxylated or 6,7-dimethoxylated), 5,6-dimethoxy-3-methyl-1-indanone, and 1-benzosuberone β-keto esters [268].

48. Condensation with Indoleethanamines

Condensation of indoleethanamines **(249)** (R = H, OMe) with cyclic β-keto esters **(250)** [R^1 = Me, X = CH_2NMe; R^1 = Et, X = CH_2, $(CH_2)_2$, $N(CH_2Ph)CH_2$] under azeotropic conditions followed by acid-catalyzed ring closure of the resulting enamines gave **(251)**.

Condensation of **(249)** with **(252)** (n = 1, R^2 = Me, n = 2, R^2 = Me, Et, OMe, CH_2OMe; n = 3, R^2 = Me) gave two types of enamines, **(253)** and **(254)**. **(253)** On treatment with acid gave alkylcyclohexanols which were further dehydrated to give cycloalkane derivatives **(255)** [269].

49. Photolytic Cleavage

Ethyl 2-oxocyclopentanecarboxylate **(55)** and ethyl 2-oxocyclohexanecarboxylate **(1)** and their 1-alkyl derivatives were photolytically cleaved to give a mixture of (E)- and (Z)-OCH(CH_2)nCH:CRCO_2Et (R= H, Me, Et,Pr,Me_2CHCH_2; n =1,2) [270[.

50. Thermal Selfcondensation

Intermoecular self-condensation of cyclic β-keto esters gave the corresponding pyran derivatives. Thus, **(256)** at 170°/24 h/12 mm Hg gave **(257)**; similarly, **(258)** gave **(259)** [271].

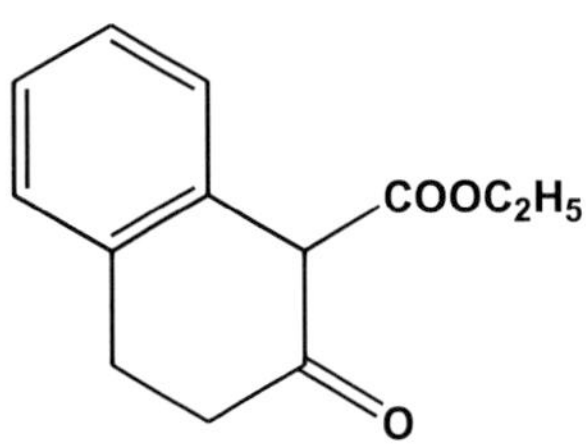

(256)

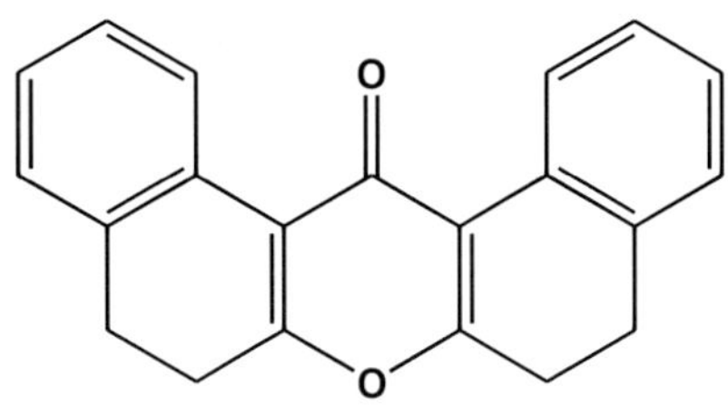

(257)

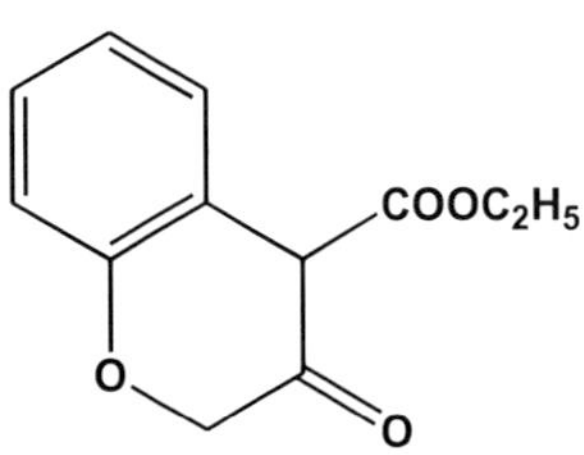

(258)

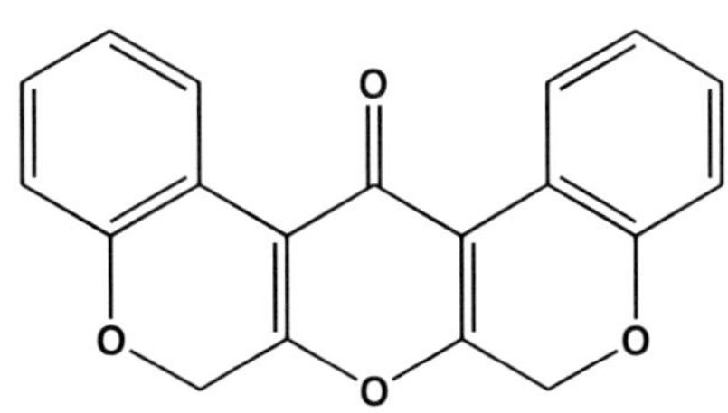

(259)

51. Condensation with Ethylene Oxide

β-Keto esters react with C_2H_4O in a way similar to ethyl acetoacetate, by replacing a mobile H atom. The reaction is one of alcoholysis resulting in breaking a ring [272].

REFERENCES

[1] R. Levine, E. Baumgarten and C.R. Hauser, J. Amer. Chem. Soc., 66, 1230 (1944).

[2] A. Kotz and A. Michels, Ann., 350, 204 (1906).

[3] W.P. Campell and M.D. Soffer, J. Amer. Chem. Soc., 64, 417 (1942).

[4] H.T. Openshaw and R. Robinson, J. Chem. Soc., 941 (1973).

[5] N.S. Vul'fson and V.I. Zaretskil, Zh. Obshch. Khim., 29, 2737 (1959); J. Gen. Chem. USSR, 29, 2704(!959); Chem. Abstr. 54, 10895 (1960).

[6] R. Tanikaga, Japan. Kagaku (Kyoto), 53(10), 52-53. (1998).

[7] S.C. Bell and S.C. Childress, J. Org. Chem. 33, 216 (1968).

[8] J.L. Luche, C. Petrier and C. Dupuy; Tetrahedron Lett., 25 (7), 753-6 (1984); Chem. Abstr., 101 (5), 3780b (1984).

[9] R.A. Bunce and C.R. Harris, J. Org. Chem., 57(25), 6981-5 (1992).

[10] R.E. Miller and F.F. Nord, J. Org. Chem. 16, 1726 (1951).

[11] B. Lal and A.P. Bhaduri, Indian J. Chem., 15B, 359 (1977).

[12] C. mannich and P. Schumann, Ber., 69 2299 (1936).

[13] P.R. Shafer, W.E. Loeb and W.S. Johanson, J. Amer. Chem. Soc., 75, 5963 (1953).

[14] A.Jr. Silveira, M. Angelastro, R. Israel, F. Totino and P. Williamsen; J. Org. Chem.; 45 (17), 3522-3, 1980; Chem. Abstr.; 93, 114010d (1980).

[15] F.W. Swamer and C.R. Hauser; J.Amer. Chem. Soc., 72, 1352-6 (1950); Chem. Abstr. 44, 6385 (1950).

[16] C.R. Hauser and B.F. Hundson; "Organic reactions" Vol. I, Ed. Roger Adams, John Wiley and Sons, inc., New York, Chapter 9, P. 266 (1942).

[17] F.W. Swamer and C.R. Hauser; J.Amer. Chem. Soc., 63, 2647 (1950).

[18] Hansely; U.S. Patent 2, 158, O71, May 16 (9139).

[19] E.S. Levin, Bull. Acad. Sci. U.R.S.S. Phys. 11, 413-18 (1947); Chem. Abstr., 42(1), 3261, (1948).
[20] E. Müller and M. Baure, Justus Liebigs Ann. Chem.; 92, 654 (1962).
[21] B. Eistert, F. Haupter and K. Schank; ibid., 665, 55 (1963).
[22] E. Wenkert and C.A. Mcpherson, J. Amer. Chem. Soc., 94, 8084 (1972).
[23] W.L. Mock and M.E. Hartmann, J. Org. Chem. 42 (3), 459 – 65 (9177); Chem. Abstr., 86, 70967 (1977).
[24] W.T. Tai and E.W. Warnhaff; Can. J. Chem., 42 (6), 1333-40; Chem. Abstr., 61, 2982 (1964).
[25] I. Shahak, Tetrahedron Letters, 20, 2201-3 (1966); Chem. Abstr., 65, 3785c (1966).
[26] S. Shunskau and K. Tadashi, Chem. Pharm. Bull., 21 (5), 1160-3 (1973); Chem., 79, 104712f. (1973).
[27] K. Kociolek, and M.T. Leplawy; Rocz. Chem., 49(11), 1841-51 (1975); Chem. Abstr., 84, 89538m (1976).
[28] R. Robinson, J. Chem. Soc., 109, 1038 (1916).
[29] G. Ehlinton and M.C. Whiting; ibid, 3052 (1953).
[30] H.O. House, R.G. Carlson, H.M. Müller, A.W. Noltes and C.D. Slater, J. Amer. Chem. Soc., 84, 2611 (1962).
[31] S. Hunig, E. Benzing and S. Lucke, Chem. Ber., 90, 2833, (1957).
[32] G. Stork, A. Brizzolara, H. Landesmann, J. Szmuszkovicz and R. Terrell, J. Amer. Chem. Soc., 85, 207-22, 1963; Chem. Abstr., 59, 474 (1963).
[33] G. Stork, R. Terrell and J. Szmuszkovicz; J. Amer. Chem. Soc., 76, 2029-30 (1954); Chem. Abstr., 49, 7538-9i (1955).
[34] O.E. Krivoshchekova and A.A. Shamshurin; Zh. Org. Khim, 7(3), 468-70 (1971); Chem. Abstr.; 75 (3) 20316j (1971).
[35] K. Nagao, M. Chiba and S. Won Kim; Synthesis 3, 197-9 (1983); Chem. Abstr., 98 (21), 178292h (1983).
[36] U. Schoellkopf, B. Banhidai, H. Frasnelli, . Meyer and H. Beckhaus, Justus Liebigs Ann. Chem., 11, 1767-83 (1974); Chem. Abstr., 82, 155230 (1975).
[37] J. Hellou, J.F. Kingston and Alex G. Fallis; Synthesis, 12, 1014-17 (1984); Chem. Abstr., 102 (23), 203635h (1985).
[38] A. Oota, (Koei Chemical Co., Ltd., Japan). Jpn. Kokai Tokkyo Koho (1998), 6 pp.
[39] H. Nishida (Mochida Pharmaceutical Co., Ltd., Japan). PCT Int. Appl., 27 pp. (2003).

[40] K. Iseki; K. Kuroki and Y. Sakamaki, (Daikin Industries, Ltd., Japan), Jpn. Kokai Tokkyo Koho, 8 pp. (2001).

[41] M.A. Metwally, E.S. Afsah and F.A. Amer, Z. Naturforsch. 36b, 1147-1148 (1981).

[42] R. Adams, Organic Reactions, Vol. 7, Chapter 3, p. 144, Wiley& Sons Inc. New York (1975).

[43] P.G. Guha, Ber., 72B, 1359 (1939).

[44] J.R. Andersen and O. Jorgensen, J. Chem. Soc. Perkin I, 3095 (1979).

[45] P. Langer; E. Holtz and N.N.R. Saleh, Chemistry-A European Journal, 8(4), 917-928(2002).

[46] D. Gravel and M. Labelle, Can. J. Chem., 63(7), 1874-83 (1985).

[47] S.J. Rhoads and R.D. Reynolds and R. Raulins, J. Am. Chem. Soc., 74 2889-92 (1952).

[48] E.J. Park; M.H. Kim and D.Y. Kim, J. Org. Chem., 69(20), 6897-6899(2004).

[49] N.T. Patil and Y. Yamamoto, Japan Journal of organic chemistry, 69(19), 6478-81(2004).

[50] U.K. Tambar and B.M. Stoltz, J. Am. Chem. Soc., 127(15), 5340-5341(2005).

[51] S.G. Will, P. Magriotis, E.R. Marinelli, J. Dolan and F. Johnson, J. Org. Chem., 50(25), 5432-3(1985).

[52] M. Kobayashi, K. Umemura, N. Watanabe and H. Matsuyama, Chemistry Letters, (7), 1067-70(1985).

[53] M. Kobayashi, K. Umemura and H. Matsuyama, Chemistry Letters, (2), 327-8(1987).

[54] G.A. Koppel and M.D. Kinnick, J. Chem. Soc.; Chem. Commun., 12, 473 (1975); Chem. Abstr. 83, 113729h (1975).

[55] H. Henecka, Ber. , 81, 197 (1984), 82, 104 (1949).

[56] D.K. Banerjee, S. Chatterijee and S.P. Bhattacharya, J. Am. Chem. Soc., 77, 411 (1955).

[57] W.S. Rapson, J. Chem. Soc., 1626 (1936).

[58] A.G. Cope and M.E. Synerholm, J. Am. Chem. Soc., 72, 5228 (1950).

[59] A.s. Dreiding and A.J. Tomasewski, J. Am. Chem. Soc., 77, 411 (1955).

[60] D. Stauffacher and H. Schinz, Helv. Chem. Acta, 37, 1223 (1954).

[61] E.M. Afsah, M.A. Metwally, F.A. Amer and M.T. El-Zimaity, Mansoura Bull. Sci., 13(2), 419-432 (1986). Z. Naturforsch. 36b, 1147-1148 (1981).

[62] K. Majima, S. Tosaki; T. Ohshima and M. Shibasaki, Tetrahedron Letters, 46(32), 5377-5381(2005).
[63] C. Simon, J.F. Peyronel, F. Clerc and J. Rodriguez, European Journal of Organic Chemistry, (19), 3359-3364(2002).
[64] G. Szollosi and M. Bartok, Chirality, 13(10), 614-618(2001).
[65] C. Simon, J.-F. Peyronel, F. Clerc and J. Rodriguez, European Journal of Organic Chemistry, (19), 3359-3364(2002).
[66] K. Shibatomi, T. Nakahashi and Y. Uozumi, Synlett, (11), 1643-1645(2000).
[67] B. Kreidler, A. Baro and J. Christoffers, Synlett, (3), 465-468(2005).
[68] K. Majima, R. Takita, A. Okada; T. Ohshima and M. Shibasaki, J. Am. Chem. Soc., 125(51), 15837-15845(2003).
[69] M. Kajiwara and K. Suzuki, (Kanto Chemical Co., Inc., Japan). Jpn. Kokai Tokkyo Koho, 6 pp (2003).
[70] M. Shibazaki, T. Oshima and S. Matsunaga, (Japan). Jpn. Kokai Tokkyo Koho, 11 pp (2002).
[71] J. Christoffers; H. Oertling. Tetrahedron, 56(10), 1339-1344(2000).
[72] M. Miesch, G. Mislin and M. Franck-Neumann, Tetrahedron Letters, 39(38), 6873-6876(1998).
[73] J.F. Lavallee, G. Berthiaume, and P. Deslongchamps, Tetrahedron Letters, 27(45), 5455-8(1986).
[74] G. Berthiaume, J.F. Lavallee and P. Deslongchamps, Tetrahedron Letters, 27(45), 5451-4(1986).
[75] R.G. Salomon and M.F. Salomon, J. Org. Chem., 40(10), 1488-92(1975).
[76] M.-H. Filippini and J. Rodriguez, Synthetic Communications, 25(2), 245-52(1995).
[77] H. Wang, M. Watanabe and T. Ikariya, Tetrahedron Letters, 46(6), 963-966(2005).
[78] J.P. Ferris, B.G. Wright and C.C. Crawford, J. Org. Chem., 30 (7), 2367-71 (1965).
[79] J. Plesek, Collection Czech. Chem. Commun., 21, 1312 (1956); 22, 49 (1957), 22, 1661 (1957).
[80] S.O. Lawesson and T. Buach, Acta Chem. Scand., 13, 1717 (1959).
[81] A. Mckillop and D.P. Rao, Synthesis, 11, 759-60 (1977); Chem. Abstr., 88, 50596m (1978).
[82] M. Pohmakotr and T. Bunlaksananusorn, Tetrahedron Letters, 41(3), 377-380(2000).
[83] C. Mannich and R. Braun, Ber., 53, 1874 (1920).

[84] M. Muhlstadt, Ber. , 93, 2638 (1960).
[85] V.P. Arya, Indian J. Chem., 14B, 982 (1976).
[86] M.A. Metwally and A.M. Khalil, J. Indian Chem. Soc. Vol. L XV, 766-767 (1988).
[87] M.A. Brimble and C. Brocke, European Journal of Organic Chemistry, (11), 2385-2396(2005).
[88] W. Wang; J. Wang and H. Li, Tetrahedron Letters, 45(39), 7243-7246(2004).
[89] T.A. Geissman and M.J. Schlatter, J. Org. chem.., 11, 771 (1946).
[90] F.M. Jaeger and L. Bijkert, Proc. Acad. Sci. Amsterdam, 40, 12 (1973); Chem. Abstr. 31, 4960 (1937).
[91] F.M. Jaeger and J.A. Van Dijk, Proc. Acad Sci. Amsterdam, 39, 384 (1936); Chem. Abstr., 30, 6341 (1936).
[92] Dieckmann and Groeneveld, Ber. , 33, 595 (1900).
[93] R.P. Linstead and A.B.L. Wang, J. Chem. Soc., 807 (1937); Chem. Abstr., 31, 5355 (1937).
[94] V.V. Feofilaktov and T.N. Ivanova, J. Gen. Chem. USSR, 13, 457 (1943); Chem. Abstr. 38, 3255 (1944).
[95] Ibid, 25, 111 (1955); Chem. Abstr., 50, 1780 (1956).
[96] H. Yasuda, Sci. Papers Inst. Phys. Chem. Research 53, 348 (1959); Chem. Abstr. 54, 15364 (1960).
[97] F.A. Amer, E.M. Afsah, M.T. El-Zimaity and M.A. Metwally, Z. Naturforsch., 34b, 95 (1979).
[98] M.A. Metwally, M.S. El-Hussiny, F.Z. El-Ablak and A.M.Khalil, Pharmazie 44 (4), 261-265 (1989).
[99] A. Baeyer, Ber. , 19, 430 (1986).
[100] H. Liebermann, Ann., 404, 272 (1914).
[101] G.P. Pendse and S. Dutt, J. Indian Chem. Soc., 9, 67 (1932).
[102] P.H. Griswold and W.S. Struve, U.S. Patent 3, 156, 719; Chem. Abstr., 62, 4031 (1965).
[103] W.S. Struve, ibid. , 2, 821, 541; Chem. Abstr., 52, 10215 (1958).
[104] E.M. Afsah, F.A. Amer, M.A. Metwally and M.R. El-Zimaity, Z. Naturforsch., 35b, 1310 (1980).
[105] F.A. Amer, E.M. Afsah, M.A. Metwally and M.T. El-Zimaity, ibid., 35b, 1395 (1980).
[106] S. Kimura, Makromol. Chem. 117, 203 (1968), Chem. Abstr., 69, 107168 (1968).110. F. Higashi, A. Tai and K. Adachi, J. Polym. Sci., 8, 2563 (1970).

[107] F. Higashi, A. Tai and K. Adachi, J. Polym. Sci., 8, 2563 (1970).108. J.A. Moore and J.E. Kochanowski, Macromolecules, 8, 121 (1975).
[108] Y. Yokoyama, K. Shibata, O. Fuji and E. Iwamoto (Japan), Yoyo Soda Kenkyu Hokoku, 19, 71 (1975); Chem. Abstr., 85, 125771 (1976).
[109] H. Fischer, W. Neuman and J.Z. Hirschbeck; Physical Chem. 1, 279 (1943).
[110] F.F. Ehrich and E.E. Jaffe, U.S. Appl., 140, 983 (1971); Chem. Abstr., 78, 99049 (1973).
[111] A. Rossi, A. Hunger, J. Kebrle and K. Hoffmann, Helv. Chim. Acta, 43, 1298 (1960).
[112] G. Ege and C. Freund, Tetrahedron lett., 39, 3677 (1978), Chem. Abstr., 90, 16855m (1979).
[113] P.M. Pihko and A. Pohjakallio, Synlett, (12), 2115-2118(2004).
[114] 115a. L. Crisci F. Broussard; C. Neri and M. Adovasio, (Great Lakes Chemical Italia Srl, Italy). PCT Int. Appl., 78 pp. 115b. Q(1998).
[115] Lan, X. Wang, R. He, C. Ding and K. Maruoka, Tetrahedron Letters 50 3280–3282(2009).
[116] L.A. Carpino and E.G.S. Rundberg, J. Org. Chem., 34, 1717 (1969).
[117] M.A. Metwally, M.S. El-Hossini, F.Z. El-Ablak and A.M. Khalil, Pharmazie, 47 (5), 336 (1992).
[118] M.H. El-Metwally, E. El-Din M. Kandeel, M.M. Abou-Elzahab, E.G. Sadek and M.A. Metwally, Croatica Chemica Acta, 65 (4), 801-811 (1992).
[119] R. Jacquier, F. Petrus, J. Verducci and Y. Vidal, Bull. Soc. Chim. France, 3664 (1971).
[120] E. Taschner, Roczinik Chem., 25, 329 (1951), Chem. Abstr. , 48, 5803 (1954).
[121] M. Botta and A.W. Dox, J. Amer. Chem. Soc., 53, 1127 (1905).
[122] H.K. Sun, Eur. Pat. Appl. Ep 123, 750 (cl Co 7 D 239/70), 07 Nov. 1984, U.S. Appl. 454, 732, 30 Dec. 1982, 18 pp., Chem. Abstr., 102 (17), 149283b (1985).
[123] J.J. Hlavaka, P. Bilha, Y.L. Lin and T. Strohmeyer, J. Heterocyclic Chem., 21 (5), 1537-41 (1984); Chem. Abstr., 102 (13), 113414f (1985).
[124] T. Tsuji and T. Ueda, Chem. Pharm. Bull., 12, 2530 (1971).
[125] B. Stanovink and M. Tisler, Croatica Chem. Acta., 44 (3), 415-18 (1972); Chem. Abstr., 78 (7), 43405P (1973).
[126] H. Antaki and V. Petrow, J. Chem. Soc., Part XII, 551 (1951); Chem. Abstr., 45, 9061 (1951).

[127] R.W. Warde, R.E. Markwell and J.B. Hughes; g. Plc. Eur Pat. Appl. Ep 193, 329 (cl CO7 D 471/O4), 03 Sep. 1986, GB Appl 85 4682, 22 Feb. 1985, 25 PP.; Chem. Abstr., 105, 22655 (1986).
[128] B. Koren, F. Kovac, A. Petric, B. Stranovinkand M. Tisler; Tetrahedron, 32 (4), 493-7, 1976; Chem Abstr., 84, 180126m (1976).
[129] V.A. Kovlunenko, V.V. Ishchenko, A.K. Tyltin, A.V. Turov and F.S. Babichev, Khim Geterotsike Soedin, 10, 1368-71 (1985); Chem. Abstr., 104 (19), 168435k (1986).
[130] W. Heinrich and L. Lulz, Chem. Ber., 111 (8), 3813-24 (1978); Chem. Abstr., 90, 22955 (1979).
[131] P. Heinz, W. Rolf and H. Guenter; Monatsh. Chem., 112 (2), 209-13 (1981), Chem. Abstr., 94, 20880k (1981).
[132] R. Adams and S.W. Mecorney, J. Amer. Chem. Soc., 66, 802 (1944).
[133] J.D. Roberts, W.T. Moreland and W. Frazer, J. Amer. Chem. Soc., 75, 637 (1953).
[134] K. Lempert, K. Simon-Omari and R. Markovits-Kornis, Acta Chim. Acad. Sci. Hung., 51, 305 (1967); Chem. Abstr., 66, 115420 (1967).
[135] J. Christoffers, J. Org. Chem., 64(20), 7668-7669(1999).
[136] A. Mittra, M. Yamashita, I. Kawasaki, H. Murai, T. Yoshioka and S. Ohta, Synlett, (8), 909-910(1997).
[137] H. Mimoun, (Firmenich SA, Switz.). PCT Int. Appl., 27 pp(2004).
[138] B.B. Snider and J.R. Duvall, Organic Letters, 6(8), 1265-1268(2004).
[139] H. Irie, J. Katakawa, M. Tomita and Y. Mizuno, Chemistry Letters, (5), 637- 40(1981).
[140] R. Helg and H. Schinz, helv. , 35, 2406 (1952).
[141] J. Sinnreich and H. Bater; Helv. , Chem. Acta 62, 156 (1979).
[142] E. Buchta and H. Bayer, Ann., 573, 227-230 (1951); Chem. Abstr. , 46, 4480 (1952).
[143] G. Bernath, G. Gondos and L. Gera, Acta Phys. Chem., 20 (1-2), 139-44 (1974); Chem. Abstr., 81, 104798e (1974).
[144] D. Buisson and R. Azerad Tetrahedron Lett. , 27 (33), 2631-4 (1986); Chem. Abstr., 106 (13), 99095b (1987).
[145] B.S. deol, D.D. Ridley and G.W. Simpson, Aug. J. CHem., 29 (11), 2459-67 (1976); Chem. Abstr., 86, 70097 (1977).
[146] K. Soai and H. Oyamada, Synthesis, 7, 605-7 (1984); Chem. Abstr., 101 (25), 229944n (1984).
[147] . C. Le Guillanton, Tetrahedron Lett., 29, 2567-70 (1972); Chem. Abstr., 90, 21953m (1979).
[148] R. Stoermer, C.W. Chydenius and E. Schinn, Ber., 578, 72 (1924).

[149] W. Steinkopf and A. Wolfram., Ann; 430, 113 (1923).

[150] A. Windaus, Ann., 447, 233 (1926).

[151] M.L. Wolform and Karabinos, J. Amer. Chem. Soc., 66, 909 (1944).

[152] M.S. Newmann and H.M. Walborsky; J. Amer. Chem. Soc., 72, 4296-7 (1950); Chem. Abstr., 45, 2867 (1951).

[153] A. Tanabe, J. Sunagawa and A. Sasaki, (Sumitomo Pharmaceutical Co., Ltd., Japan). Jpn. Kokai Tokkyo Koho, 17 pp(2004).

[154] C.A.M. Fraga, L.H.P. Teixeira; C. M. de S. Menezes, C.M.R. Sant'Anna, M.da C.K.V. Ramos, F.R. de Aquino Neto and E.J. Barreiro, Tetrahedron, 60(12), 2745-2755(2004).

[155] S. Danchet; D. Buisson and R. Azerad, Journal of Molecular Catalysis B: Enzymatic, 5(1-4), 255-259(1998).

[156] C. Abalain, D. Buisson and R. Azerad, Tetrahedron: Asymmetry, 7(10), 2983-2996(1996).

[157] .S. Yadav, S. Nanda, P.T. Reddy and A.B. Rao, J. Org. Chem., 67(11), 3900-3903(2002).

[158] D. Buisson, X. Baucherel, E. Levoirier and S. Juge, Tetrahedron Letters, 41(9), 1389-1392(2000).

[159] J.P. Genet, X. Pfister, V. Ratovelomanana-Vidal; C. Pinel and J.A. Laffitte, Tetrahedron Letters, 35(26), 4559-62, 159b(1994). S. K. Padhi,_ I. A. Kaluzna, D. Buisson, R. Azeradb and J. D. Stewarta, Tetrahedron: Asymmetry 18, 2133–2138(2007).

[160] S. Sethna and R. Phadke, Org. Reactions, 7, 1-58 (1953).

[161] R. Adams and C.M. Smith and S. Loewe, J. Amer. Chem. Soc., 63, 1971 (1941).

[162] R.D. Desai, M.M. Gaitonde, H.S. Mehdi and R. Shah, Proc. Indian Acad. Sci., 25 (A), 345 (1947); Chem. Abstr., 42, 1913 (1948).

[163] R. Ghosh, A.R. Todd and S. Wilkinson, J. Chem. Soc. PartII, 1121 (1940).

[164] H.K. Sen and U.P. Basu, J. Indian Chem. Soc., 5, 467 (1928); Chem. Abstr., 23, 383 (1929).

[165] H.V. Pechmann and B. Duisberg; Ber. Dtsch. Chem. Bis., 16, 2115 (1983).

[166] N.A. Chowdhry and R.D. Desai, Proc. Indian Acad. Sci., 8A, 12 (1938); Chem. Abstr., 32, 9066 (1938).

[167] R. Verhe, N. Schamp and J. Van Damme; Rijksuniv. Gent. 38 (3), 510-18 (1973); Chem. Abstr., 81, 49519n (1974).

[168] R. Verhe, P. Chielens and N. Schamp; Rijksuniv. Gent. 38 (1), 119-24 (1973); Chem. Abstr., 80, 13362t (1974).

[169] N.H. Pardanani, M.G. Parekh and K.N. Trivedi, J. Indian Chem. Soc., 46, 1014 (1969).

[170] N.H. Pardanani and K.N. Trivedi, J. Institution of Chemists, 42, 153 (1970).

[171] V.N. Dholakia and K. N. Trivedi, J. Indian Chem. Soc., 47, 1059 (1970).

[172] N.H. Pardanani, and K.N. Trivedi, J. Indian Chem. Soc., 48, 339(1971).

[173] P.N. Confalone and D.L. Confalone, Tetrahedron, 39 (8), 1265-71 (1983); Chem. Abstr., 99 (7), 53638q (1983).

[174] Y.A. Shaikh and K.N. Trivedi, J. Indian Chem. Soc., 49, 877 (1972).

[175] Ibid. J. Indian Chem. Soc., 51 (8), 755-6 (1974); Chem. Abstr., 82, 170744f (1975).

[176] H.K. Sen and U.P. Basu, J. Indian Chem. Soc., 5, 470 (1928).

[177] M.G. Parekh and K.N. Trivedi, Aust. J. Chem., 23, 407 (1970).

[178] K.P. Shah and K.N. Trivedi, Aust. J. Chem., 27 (29), 1971-6 (1974); Chem. Abstr., 81, 152053e (1974).

[179] T. Kappe and C. Mayer, Synthesis, 7, 524-6 (1981); Chem. Abstr., 95, 150496s, 1981.

[180] O.S. Wolfbeis, Montash Chem., 113(3), 365-70, 982; Chem. Abstr. 97 (4); 215, 193t, (1982).

[181] J.R. Merchant, N.M. Koshti and K.M. Bakre; J. Heterocyclic Chem., 18(8), 1655-8, (1981); Chem. Abstr., 96 (19), 162487y, (1982).

[182] K. Chebaane, M. Guyot and D. Molho, Bull. Soc. Chem. Fr., 11-12 (pt.2), 2516-20, (1975) Chem. Abstr. , 84, 179968f, (1976).

[183] F. Lauria, V. Vecchietti, R. Tommansini and N. Passerini, Ger. Offen. 2, 459, 076 (C1 C07 D, A61 K) 10 Jul 1976, Ital Appl. 32, 088173, 27 Dec 1973, 42 PP; Chem. Abstr., 83, 163996g (1975).

[184] C. Mentzer, D. Molho and P. Vercier, Bull. Soc. Chem. France, 91-3, 1952; Chem. Abstr., 47, 4336f (1953).

[185] C. Mentzer, D. Molho and P. Vercier, Compt. Rend., 232, 1488-99, 1951; Chem. Abstr., 46, 4003d (1952).

[186] Y.S. Agasimundin and S. Rajagopal, J. Org. Chem., 30, 2084 (1967).

[187] Y.S. Agasimundin and S. Rajagopal, Chem. Ber., 100, 383, (1967).

[188] J.S.H. Devies, Lambf. and Suschitzky H., J. Chem. Soc., 2, 1958 (1970).

[189] A. Mustafa Sidky, M.M. S.M.A. Sayer and Solimann F.M., Tetrahedron, 19, 1335 (1963).

[190] F. Scheimann and H. Suschilzky, Tetrahedron, 7, 31 (1959).

[191] K.D. Kaufman and L.R. Worden, J. Org. Cdhem., 26, 4721, (1961).

[192] K.P. Sanghvi, R.J. Patolia and K.N. Trivedi, J. Indian Chem. Soc. 56 (1), 52, 5 (1979).

[193] W. Untze, Fette, Seifen, Anstichm., 75 (6), 393-4, 1973; Chem. Abstr., 79, 126235e (1973).

[194] J. N. Chatterjea, S.C. Shaw and N.D., J. Indian Chem. Soc., 51 (8), 752-4, 1974; Chem. Abstr. 82, 125222y (1975).

[195] K.M. Bakre and J.R. Merchant, Indian J. Chem., 20B (4), 346-7, 1981; Chem. Abstr., 95, 97518x (1981).

[196] M.A. Metwally and A.M. Khalil, J. Indian Chem. Soc. Vol. LXV, 766-767 (1988).

[197] E.M. Afsah, M.A. Metwally, F.A. Amer and M.T. El-Zimaity; Mansoura Sci. Bull. 13 92), 419 (1986).

[198] B.P. Choudhari and V.V. Mulwad. Indian Journal of Chemistry, Section B:, 42B(9), 2080-2085(2003).

[199] Mitsui Toatsu Chemicals, Inc. Jpn. Kokai Tokkyo Koho J.P. 57, 183, 744, 12 Nov. 1982, Appl. 81/66, 881, 06 May 1981; Chem. Abstr., 98, 125631 (1981).

[200] . C.W.T. Hussey and A.R. Pinder, J. Chem. Soc. Part 1, 1517-18, 1962; Chem. Abstr. 57, 2089 (1962).

[201] C.J. Sheehan and E.C. Mumaw, J. Amer. Chem. Soc., 72, 2127 (1950).

[202] A. Koltz, Annaler, 358, 183, 1908, Kotz and Grethe; J. Prakt. Chem., 80, 473 (1909).

[203] H. Ruhkopf, Ber., 72B, 1978, 1939; Chem. Abstr. 34, 760 (1940).

[204] M. Mousseron, R. Jacquier, A. Fontaine and R. Zagdoum; Bull. Soc. Chim. France, 1246, 1954; Chem. Abstr. 49, 139/ 8h (1945).

[205] J.E. Brenner, J. Org. Chem., 26, 22 (1961).

[206] . I. Shotta, S. Norio and H.A. Kiyoloka; Yukagaku, T., 29 (I), 270-1, 1980, Chem. Abstr.., 93 (11), 114012 (1980). 206b. S. Suzuki, H. Furuno, Y. Yokoyamac and J. Inanagab, Tetrahedron: Asymmetry 17 (2006) 504–507

[207] G. Wittig, H. Eggers and P. Duffner; Justus Libigs Ann. Chem., 10, 619, 1958; Griffin and Wilschard; J. Org. Chem., 28, 1715, 1963; Schweizer; J. Am. Chem. Soc., 86, 2744, (1964).

[208] K.B. Becker, Helv. Chem. Acta, 60 (1), 60-80, 1977; Chem. Abstr., 86, 170927 (1977).

[209] C.A. Greene, A. Cruz and P. Crabbe, Tetrahedron Lett. , 31, 2707-8, 1976, Chem. Abstr., 86, 55021p (1977).

[210] S.B. Huang, E.J. Parish and O.M. Howard, J. Org. Chem., 3-9 (17), 2647, 1974, Chem. Abstr., 81, 104812e (1974).

[211] A.P. Krapcho and A.J. Lovey, Tetrahedron lett. , 12, 957-60, 1973, Chem. Abstr., 79, 17996h (1973).
[212] A.P. Krapcho, J.F. Weimaster, J.M. El Clridge, E.C.E. Jahngen, Jr. A.J. Lovey and W.P. Slephens; J. Org. Chem., 43 (1), 138-47, 1978; Chem. Abstr. 88, 50032 Z (1978).
[213] A.P. Krapcho, E.C.E. Jahngen, Jr. A.J. Lovey J.F. and F.W. Short, Tetrahedron lett., 13, 1091-4, 1974; Chem. Abstr. 81, 49072m (1974).
[214] Y. Tsuda and Y. Sakai; Synthesis, 2, 119-20, 1981; C.A., 95, 61906h (1981).
[215] D. Paquer, S. Samadja, J. Vialle and C.R. Hebd. Seances Acad. Sci., Ser. C. 279 (12), 529-31, 1974; Chem. Abstr.; 82, 170144k (1975).
[216] F. Lions; J. Proc. Roy. Soc. N.S. Wales, 66, 516-25, 1933, Chem. Abstr., 27, 2954 (1933).
[217] A. De Groot and B.J.M. Jansen, Rect trav. Chim. Pays- bas, 98 (9), 487-92, 1979, Chem. Abstr., 92, 41495h (1980).
[218] J.L. Van Der. Bann. And F. Bickelhaupt, Chem. Comm., 326 (1971).
[219] F. Brody and P.R. Ruby; "In Pyridine and its derivatives" Edited E. Klingsberg Part 1, Chapter II Inter. Science, New York (1960).
[220] H. Meislich, Ibid, Part 3, Chapter XII (1962).
[221] J.L. Van Der. Bann. And F. Bickelhaupt, Tetrahedron, 30 (15), 2447-53, 1974; Chem. Abstr., 82, 43141g (1975).
[222] T. Sugawara and Kuwajima; Tetrahedron Letters, 26 945), 5571-4, 1984; Chem. Abstr., 105, 6204C (1986).
[223] M. Sugi, D. Sakuma and H. Togo. J. Org. Chem., 68(20), 7629-7633(2003).
[224] M. Sugi and H. Togo, Tetrahedron, 58(16), 3171-3175(2002).
[225] E. Hasegawa; T. Kitazume; K. Suzuki and E. Tosaka, Tetrahedron Letters, 39(23), 4059-4062(1998).
[226] S.H. Chung, M.S. Cho, J.Y. Choi, D.W. Kwon and Y.H. Kim, Synlett, (8), 1266-1268(2001).
[227] R.S. Atkinson, E. Barker; P.J. Edwards and G.A. Thomson, J. Chem. Soc., Perkin Transactions 1, (12), 1533-42(1995).
[228] P. Dowd and S.C. Choi, Tetrahedron, 48(23), 4773-92(1992).
[229] Z.F. Xie and K. Sakai, J. Org. Chem., 55(3), 820-6(1990).
[230] A.J. Frew and G.R. Proctor, J. Chem. Soc., Perkin Transactions 1: (1972-1999), (6), 1245-50 (1980).
[231] F.D. Sancilio and J.F. Blount, Acta Crystallographica, Section B: Structural Crystallography and Crystal Chemistry, B32 (7), 2123-7(1976).

[232] D.W. Adamson, J. Chem. Soc., 1564 (1939).

[233] W.E. Noland, R.J. Sundberg and M.L. Michaelson, J. Org. Chem., 28 (12), 3576-7, 1963; Chem. Abstr., 60, 5481 (1964).

[234] D. Nwoko, M. Wells and L. Bock, (Crompton Corporation, USA). U.S8 pp (2002),

[235] J. Sinnreich and H. Batzer, (Ciba-Geigy A-G) Ger. Offen. 2, 437, 875 (CI. C07C, C08G), 20 Feb. 1975, Swiss Appt. 11, 512173, 09 Aug. 1973; 11 PP., Chem. Abstr., 83, 11133 (1975).

[236] M.R. Acocella, O.G. Mancheno, M. Bella and K.A. Jorgensen, J. Org. Chem. , 69(23), 8165-8167(2004).

[237] X. Baucherel, E. Levoirier, J. Uziel and S. Juge, Tetrahedron Letters, 41(9), 1385-1387(2000).

[238] M. Marigo, N. Kumaragurubaran and K.A. Jorgensen, Chemistry--A European Journal, 10(9), 2133-2137(2004).

[239] B. Das, K. Venkateswarlu, G. Mahender and I. Mahender, Tetrahedron Letters, 46(17), 3041-3044(2005).

[240] H.M. Meshram, P.N. Reddy, K. Sadashiv and J.S. Yadav, Tetrahedron Letters, 46(4), 623-626(2005).

[241] J.A. Ma and D. Cahard, Tetrahedron: Asymmetry, 15(6), 1007-1011 (2004).

[242] M. Sodeoka and Yo. Hamashima. (Takasago Perfumery Co., Ltd., Japan). Jpn. Kokai Tokkyo Koho, 30 pp(2004).

[243] H. Yamanaka, K. Hisaki, K. Kase, T. Ishihara and J.T. Gupton, Tetrahedron Letters (, 39(24), 4355-43581998).

[244] M. Sasidharan and R. Kumar, Journal of Molecular Catalysis A: Chemical, 210(1-2), 93-98(2004).

[245] J. Khiari, H.B. Ben and D. Gravel, Tunisia. Physical & Chemical News, 7(1), 130-133(2002).

[246] J. Khiari and D. Gravel, Synthetic Communications, 29(11), 1929-1932(1999).

[247] O.V. Gulyakevich, A.A. Govorova, P.V. Kurman, A.S. Lyakhov and A.L. Mikhal'chuk, Chemistry of Heterocyclic Compounds (NY), 40(10), 1288-1294(2004).

[248] F.M.A. El-Taweel and T.M.Abu. Elmaati, Journal of the Chinese Chemical Society (Taipei, Taiwan), 49(6), 1051-1055(2002).

[249] S. Li, R.K. Dieter and L. Howard, J. Org. Chem., 68(3), 969-973(2003).

[250] A. Rojas, A. Perez-Encabo, I. Herraiz-Sierra, and C.A. Blanco, Can. J. Chem., 79(4), 448-454(2001).

[251] S.I. Yakimovich, V.A. Khrustalev and L.N. Ignatyuk, Zhurnal Organicheskoi Khimii, 14(7), 1456-62(1978).
[252] M.H. Filippini and J. Rodriguez, J.Chem. Soc., Chemical Communications, (1), 33-4. 253(1995).
[253] M. Miesch, G. Mislin and M. Franck-Neumann, Tetrahedron Letters, 39(38), 6873-6876(1998).
[254] F. Yoshizako, M. Ogino, A. Nishimura, M. Chubachi and T. Horii, Journal of Fermentation and Bioengineering, 79(2), 141-5(1995).
[255] J. Montalt, F. Linker, F. Ratel and M. Miesch, J. Org. Chem., 69(20), 6715-6721(2004).
[256] D. Yang, J.H. Li, Q. Gao and Y.L. Yan, Organic Letters, 5(16), 2869-2871 (2003).
[257] A. Wolfson, I.F.J. Vankelecom, S. Geresh and P.A. Jacobs, Journal of Molecular Catalysis A: Chemical, 198(1-2), 39-45(2003).
[258] N. Desroy; R. Le Roux; P. Phansavath; L. Chiummiento; C. Bonini and J.P. Genet, Tetrahedron Letters, 44(9), 1763-1766(2003).
[259] G. Wittmann; G. Gondos and M. Bartok, Helvetica Chimica Acta (1990), 73(3), 635-9.
[260] L. Wei and W.D. Lubell, Can. J. Chem., 79(1), 94-104(2001).
[261] K. Van Emelen, T. De Wit, G.J. Hoornaert and F. Compernolle, Organic Letters, 2(20), 3083-3086(2000).
[262] T. Tanaka, T. Azuma, X. Fang, S. Uchida; C. Iwata, T. Ishida, Y. In and N. Maezaki, Synlett, (1), 33-36(2000).
[263] A. De Groot and B.J.M. Jansen, Recueil des Travaux Chimiques des Pays-Bas, 98(9), 487-92(1979).
[264] F. Bellina, D. Ciucci, R. Rossi and P. Vergamini, Tetrahedron, 55(7), 2103-2112(1999).
[265] K. Kato, H. Suemune and K. Sakai, Heterocycles, 37(1), 413-23(1994).
[266] A.G. Schultz and M.A. Holoboski, Tetrahedron Letters, 34(19), 3021-4(1993).
[267] L.A. Paquette, K. Dahnke, J. Doyon, W. He; K. Wyant and D. Friedrich, J. Org. Chem., 56(21), 6199-205(1991).
[268] O. Duval and L.M. Gomes, Tetrahedron, 45(14), 4471-6(1989).
[269] G. Bobowski ands J. Jr. Shavel, J. Heterocyclic Chem., 22(6), 1679-88(1985).
[270] J. Kossanyi, J. Perales, A. Laachach, I. Kawenoki and J.P. Morizur, Synthesis, (4), 279-81(1979).
[271] R. Verhe, L. De Buyck, N. De Kimpe and N. Schamp, Bulletin des Societes Chimiques Belges, 87(1), 83-4(1978).

[272] K.G. Pakendorf and F.F. Machus. Doklady Akademii Nauk SSSR, Seriya A, 31 441-3(1941).

INDEX

A

B

N

O

P

Q

R

S

T

U

W

X

Y

Z